Ian Dempsey, Michael O'Neill and Anthony Brabazon

Foundations in Grammatical Evolution for Dynamic Environments

Studies in Computational Intelligence, Volume 194

Editor-in-Chief
Prof. Janusz Kacprzyk
Systems Research Institute
Polish Academy of Sciences
ul. Newelska 6
01-447 Warsaw
Poland
E-mail: kacprzyk@ibspan.waw.pl

Further volumes of this series can be found on our homepage: springer.com

Vol. 171. Chi-Keong Goh, Yew-Soon Ong and Kay Chen Tan (Eds.)
Multi-Objective Memetic Algorithms, 2009
ISBN 978-3-540-88050-9

Vol. 172. I-Hsien Ting and Hui-Ju Wu (Eds.)
Web Mining Applications in E-Commerce and E-Services, 2009
ISBN 978-3-540-88080-6

Vol. 173. Tobias Grosche
Computational Intelligence in Integrated Airline Scheduling, 2009
ISBN 978-3-540-89886-3

Vol. 174. Ajith Abraham, Rafael Falcón and Rafael Bello (Eds.)
Rough Set Theory: A True Landmark in Data Analysis, 2009
ISBN 978-3-540-89886-3

Vol. 175. Godfrey C. Onwubolu and Donald Davendra (Eds.)
Differential Evolution: A Handbook for Global Permutation-Based Combinatorial Optimization, 2009
ISBN 978-3-540-92150-9

Vol. 176. Beniamino Murgante, Giuseppe Borruso and Alessandra Lapucci (Eds.)
Geocomputation and Urban Planning, 2009
ISBN 978-3-540-89929-7

Vol. 177. Dikai Liu, Lingfeng Wang and Kay Chen Tan (Eds.)
Design and Control of Intelligent Robotic Systems, 2009
ISBN 978-3-540-89932-7

Vol. 178. Swagatam Das, Ajith Abraham and Amit Konar
Metaheuristic Clustering, 2009
ISBN 978-3-540-92172-1

Vol. 179. Mircea Gh. Negoita and Sorin Hintea
Bio-Inspired Technologies for the Hardware of Adaptive Systems, 2009
ISBN 978-3-540-76994-1

Vol. 180. Wojciech Mitkowski and Janusz Kacprzyk (Eds.)
Modelling Dynamics in Processes and Systems, 2009
ISBN 978-3-540-92202-5

Vol. 181. Georgios Miaoulis and Dimitri Plemenos (Eds.)
Intelligent Scene Modelling Information Systems, 2009
ISBN 978-3-540-92901-7

Vol. 182. Andrzej Bargiela and Witold Pedrycz (Eds.)
Human-Centric Information Processing Through Granular Modelling, 2009
ISBN 978-3-540-92915-4

Vol. 183. Marco A.C. Pacheco and Marley M.B.R. Vellasco (Eds.)
Intelligent Systems in Oil Field Development under Uncertainty, 2009
ISBN 978-3-540-92999-4

Vol. 184. Ljupco Kocarev, Zbigniew Galias and Shiguo Lian (Eds.)
Intelligent Computing Based on Chaos, 2009
ISBN 978-3-540-95971-7

Vol. 185. Anthony Brabazon and Michael O'Neill (Eds.)
Natural Computing in Computational Finance, 2009
ISBN 978-3-540-95973-1

Vol. 186. Chi-Keong Goh and Kay Chen Tan
Evolutionary Multi-objective Optimization in Uncertain Environments, 2009
ISBN 978-3-540-95975-5

Vol. 187. Mitsuo Gen, David Green, Osamu Katai, Bob McKay, Akira Namatame, Ruhul A. Sarker and Byoung-Tak Zhang (Eds.)
Intelligent and Evolutionary Systems, 2009
ISBN 978-3-540-95977-9

Vol. 188. Agustín Gutiérrez and Santiago Marco (Eds.)
Biologically Inspired Signal Processing for Chemical Sensing, 2009
ISBN 978-3-642-00175-8

Vol. 189. Sally McClean, Peter Millard, Elia El-Darzi and Chris Nugent (Eds.)
Intelligent Patient Management, 2009
ISBN 978-3-642-00178-9

Vol. 190. K.R. Venugopal, K.G. Srinivasa and L.M. Patnaik
Soft Computing for Data Mining Applications, 2009
ISBN 978-3-642-00192-5

Vol. 191. Zong Woo Geem (Ed.)
Music-Inspired Harmony Search Algorithm, 2009
ISBN 978-3-642-00184-0

Vol. 192. Agus Budiyono, Bambang Riyanto and Endra Joelianto (Eds.)
Intelligent Unmanned Systems: Theory and Applications, 2009
ISBN 978-3-642-00263-2

Vol. 193. Raymond Chiong (Ed.)
Nature-Inspired Algorithms for Optimisation, 2009
ISBN 978-3-642-00266-3

Vol. 194. Ian Dempsey, Michael O'Neill and Anthony Brabazon
Foundations in Grammatical Evolution for Dynamic Environments, 2009
ISBN 978-3-642-00313-4

Ian Dempsey
Michael O'Neill
Anthony Brabazon

Foundations in Grammatical Evolution for Dynamic Environments

Springer

Dr. Ian Dempsey
Lead Research Analyst
Pipeline Financial Group, Inc.
60 East 42nd St. Suite 624
New York, NY 10165
USA
E-mail: ian.dempsey@pipelinefinancial.com

Prof. Anthony Brabazon
Head of Research - School of Business
Quinn School
University College Dublin
Belfield, Dublin 4
Ireland
E-mail: anthony.brabazon@ucd.ie

Dr. Michael O'Neill
Director - Natural Computing Research
and Applications Group
School of Computer Science and
Informatics
University College Dublin
Belfield, Dublin 4
Ireland
E-mail: m.oneill@ucd.ie

ISBN 978-3-642-00313-4 e-ISBN 978-3-642-00314-1

DOI 10.1007/978-3-642-00314-1

Studies in Computational Intelligence ISSN 1860949X

Library of Congress Control Number: 2009920696

Typeset & Cover Design: Scientific Publishing Services Pvt. Ltd., Chennai, India.

Printed in acid-free paper

9 8 7 6 5 4 3 2 1

springer.com

Preface

Dynamic environments abound and offer particular challenges for all optimisation and problem solving methods. A well-known strategy for survival in dynamic environments is to adopt a population-based approach. Rather than maintaining a single candidate solution, a population of candidate solutions is employed. This allows a diversity of potential solutions to be maintained, which increases the likelihood that a sufficient solution exists at any point in time to ensure the survival of the population in the long term. Dynamic environments can exhibit different types of change that can be abrupt and random, cyclical, or the product of complex relationships. The changes might range from relatively small smooth transitions to substantial perturbations in all aspects of the domain.

Natural Computing (NC) has given rise to a family of population-based algorithms that exhibit varying degrees of success in solving problems in dynamic environments. It is natural to turn to algorithms which are inspired by the natural world when one wishes to solve problems in the natural world. In particular, biological evolution has given rise to effective problem solvers which survive in complex dynamic environments. Without natural evolution, the inspriation for evolutionary compuation, we would not have any of the other NC algorithms such as neurocomputing, immunocomputing, sociocomputing and grammatical and developmental computing; they are inspired by the products of the biological evolutionary process acting in a dynamic environment.

In this book we focus on the first steps in the extension of a grammar-based form of Genetic Programming, Grammatical Evolution, in order to improve its ability to solve problems in dynamic environments. A relatively recent, powerful, addition to the stable of Evolutionary Computation, Grammatical Evolution (GE) adopts BNF grammars for the evolution of variable length programs. Thus far, there has been little study of the utility of GE in dynamic environments. **Foundations in Grammatical Evolution for Dynamic Environments** is the second book to be published on Grammatical Evolution, and it has been six years since *Grammatical Evolution:*

Evolutionary Automatic Programming in an Arbitrary Language appeared. A comprehensive analysis of prior work in EC and GE in the context of dynamic environments is presented. From this, it is seen that GE offers substantial potential due to the flexibility provided by the BNF grammar and the many-to-one genotype-to-phenotype mapping.

Subsequently, novel methods of constant creation are introduced that incorporate greater levels of latent evolvability through the use of BNF grammars. These methods are demonstrated to be more accurate and adaptable than the standard methods adopted.

Through placing GE in the context of a dynamic real-world problem, the trading of financial indices, phenotypic diversity is demonstrated to be a function of the fitness landscape. That is, phenotypic entropy fluctuates with the universe of potentially fit solutions. Evidence is also presented of the evolution of robust solutions that provide superior out-of-sample performance over a statically trained population.

The findings in this study highlight the importance of the genotype-to-phenotype mapping for evolution in dynamic environments and uncover some of the potential benefits of the incorporation of BNF grammars in GE.

New York & Dublin,
January 2009

Ian Dempsey
Michael O'Neill
Anthony Brabazon

Acknowledgements

Writing a book is a task not set upon lightly, and without the encouragement and support of many individuals this book would not have been possible. In particular, we are most grateful for our parents (clearly best-of-generation individuals) who are always a constant well of encouragement. Ian has special thanks to friends who still call despite weeks of absence. Michael is especially grateful for the patience and continuous support of his family, Gráinne, Aoife and Michael J., who often sacrifice family time for writing. Tony thanks Maria for her unending support.

We would like to thank the members (past and present) of the UCD Natural Computing Research & Applications Group (NCRA) for the countless debates and invaluable insights into the topics presented in this book. The recent results on digit concatenation and problem difficulty were made possible through the curiosty and hard work of Jonathan Byrne, Erik Hemberg and James McDermott. Our thanks also extends to the supportive environment created in the UCD Complex & Adaptive Systems Laboratory, which is home to the NCRA, and to all the staff of the UCD School of Computer Science & Informatics and the UCD School of Business. Without the support and nurturing environments of our Schools and CASL this book could not have been realised. Special thanks to Chris Stephens for an engaging discussion on the taxonomy of change and efficient markets. The insights gained contributed significantly to this book.

Finally we also extend our thanks to Dr. Thomas Ditzinger of Springer-Verlag and to Professor Janusz Kacprzyk, editor of this book series, for their encouragement of, and their support during, the preparation of this book.

To My Parents
ID

To Gráinne, Aoife and Michael J.
MON

To Maria
AB

Contents

Chapter 1
Introduction

The biological organisms that populate our planet today are widely considered as the product of an evolutionary process. A process that over the course of time, allowed the adaptation of the ancestors of these organisms so that they were better enabled to survive in their environment. Coupled with this passage of time, the world and the dynamics of the predators and prey of these organisms have also transformed and evolved, resulting in an environment that is permanently undergoing change. Natural evolution has enabled a rich and diverse range of organisms to survive and prosper under these circumstances. It is this process that is the inspiration behind the field of Evolutionary Computation (EC). However, despite natural evolution being set in a fundamentally dynamic environment, the majority of research in EC has been dedicated to overcoming issues encountered in solving static problems and optimising algorithms for these problems. Considering that most real-world problems, like biological organisms, are set in dynamic environments and that these algorithms face issues like premature convergence when operating under such circumstances, the long term future and mainstream adoption of EC is jeopardised unless the issues involved in conducting evolution in dynamic environments are identified and addressed.

Given the inspiration behind EC, it seems the greatest potential contribution of EC is in its application to problems set in dynamic or time-varying environments. In order to achieve this, weaknesses in the current state of the art must be understood, and efficient and effective mechanisms developed to bring the capability of EC more in line with its natural inspiration. This book categorises the types of changes faced in dynamic environments and identifies features of algorithms that are advantageous under these conditions. The dynamic environment of financial markets is explored and novel features are developed and analysed to aid in the evolutionary process.

In this book we present the first steps in the extension of a grammar-based form of Genetic Programming (GP), Grammatical Evolution (GE), in order to improve its ability to solve problems in dynamic environments. We also wish to better understand the behaviour of other GP-based paradigms under

I. Dempsey et al.: Foundations in Gramm. Evol. for Dyn. Environ., SCI 194, pp. 1–8.
springerlink.com

dynamic environments. Given the power of the biological process of evolution to adapt to ever changing environments, it is surprising that the number of studies applying and explicitly studying their artificial counterpart of GP in dynamic environments have been minimal [54]. While some applications in dynamic environments have been undertaken, largely in the past two years, (e.g., [229, 217, 241, 84, 97, 107, 92]), there has been little analysis of the behaviour of GP in these environments. The two main examples have examined bloat [116] and constant generation [54]. An important contribution of this book is to begin to address this large and important research gap.

The rest of this chapter is structured as follows: the next section offers an overview of dynamic problems. An introduction to the EC paradigm and the area focused on in this study, GE is then given. Following from this an overview of the structure of the remainder of this book then completes the chapter.

1.1 Dynamic Problems

Generally where static problems are concerned the problem setting is defined a-priori and the system evolves its population until a deterministic solution or a solution that satisfies some terminating criteria is found. The population pool is optimised for the problem at hand. With dynamic problems however, as time progresses, some feature of the environment or preferences of the fitness function may change, necessitating a period of retraining. In this respect, dynamic problems are time-varying; the fitness landscape alters as a function of time. This variation with time is a feature that is manifested in many real-world problems and presents significant opportunities for the effective application of EC.

In dealing with dynamic environments, the researcher is faced with two options when some element has changed: to either restart afresh with a new population, or to allow the existing population to continue to adapt to the new environment. If the change that has occurred is not a radical departure from the previous landscape, it stands to reason that adapting the current population represents a more efficient approach. In this case, because solutions are evolved over varying landscapes, they tend to be more robust as opposed to the specific optimisation that occurs with static problems. For dynamic environments the focus shifts to that of *survival* in a changing environment from that of attaining an optimal or perfect solution in a static environment.

With this shift in focus, the domain of dynamic environments carries with it a number of challenges for EC. Part of the strength of Genetic Algorithms (GAs) is their ability to conduct a wide search of the solution landscape. However, once an optimal solution is found the population is seen to converge and lose diversity thus reducing the algorithm's ability to discover a new solution when the environment changes. Does the algorithm require a means of maintaining or generating diversity within the population? Or would a means for partitioning the population into separate sub-populations provide a better way to explore discrete areas of the search space?

There are also questions as to whether mechanisms should be introduced to provide the algorithm with a form of memory so that it can retrieve past effective solutions or re-introduce previously good genetic material. Or does such a mechanism render the efficiency gain in adapting an evolved population to a new change redundant, if the extra computational cost of maintaining this mechanism is taken into account? Instead does a population-based memory or competency suffice? Is the population able to encode fundamental information about the problem that can help it adapt to future changes?

More fundamentally a number of different types of dynamic problems are observed. Different problems may have different constraints (static or variable), and may involve different kinds of change, which could vary with respect to the severity and rate of change that might be possible.

Dynamic problems also present challenges to EC in the basic area of algorithm-performance analysis. In static environments, the end result of the algorithm measured against the goal presents a quantitative and qualitative perspective of its performance. It also allows for comprehensive meaningful comparative analysis with other algorithms tested on the same problem. When the problem varies as a function of time, however, a number of extra and different dimensions come into play as the goal of the algorithm is to track the optima not "solve" a problem. The final fitness of the best solution produced at any time only represents one snapshot of the algorithm's performance over time. As the conditions have changed over the course of the run, the potential level of difficulty may also have varied. The performance of the population over time must be monitored in terms of its ability to consistently produce a good fitness and its ability to adapt to changes and minimise major errors.

The development of algorithms that can cope with these issues is crucial to the long term future of the field and its utility in the real-world. A more detailed examination of dynamic environments and the methods used to deal with them is covered in Chapter 3.

1.2 EC and Grammatical Evolution

Considering the success of natural evolution in producing solutions to the myriad of problems faced by living organisms, as outlined above, computer scientists have adopted many of the empirically observed mechanisms used in nature as they develop algorithms to solve the challenges they encounter.

The first example of the concept of adopting an evolutionary search to afford computers the ability to learn can be traced back to the founder of the field of Artificial Intelligence, Alan Turing. In 1948 he suggested that [219]:

"There is the genetical or evolutionary search by which a combination of genes is looked for, the criterion being the survival value."

Then in 1950 he followed this thinking with concepts of *selection, hereditary material,* and *mutation* where he noted [220]:

> *"We cannot expect to find a good child-machine at the first attempt. One must experiment with teaching one such machine and see how well it learns. One can then try another and see if it is better or worse. There is an obvious connection between this process and evolution, by the identifications*
> *"Structure of the child machine = Hereditary material"*
> *"Changes of the child machine = Mutations"*
> *"Natural selection = Judgment of the experimenter"*

Thus, Turing identified some of the key concepts of EC. Later in the 1950s Friedberg developed his "Learning Machine" that applied mutations to assembly code, though no notion of selection was used [71, 72]. In 1966 Fogel [68] reached another milestone with the evolution of finite state machines.

The GA was introduced by John Holland in the 1970s [90]. In Holland's GA he described the notion of a population of fixed-length solutions that are initially randomly generated and then tested against the problem at hand. The best performing individuals are selected for reproduction, with their offspring created through the use of genetic operators such as crossover and mutation, then reintroduced to the population replacing the weaker solutions. This represents the evolution of one generation. Subsequent generations are created until some termination criteria is met.

The first evidence of the evolution of non-assembly programs was presented by Cramer in 1985 [46]. His aim was to use Holland's GA to evolve syntactically correct programs. Cramer followed two strands of representation for his programs: linear and tree-based representations. While the linear representation encountered little success in producing syntactically correct programs, the tree based representation did using sub-tree crossover to produce the compliant child programs. In quick succession this work was followed by both Hicklin [88] and Fujiki [73] in 1986, who were first to describe the use of grammars for the generation of variable-length program structures, and by Dickmanns and Schmidhuber in 1987 [58] who used a GA to evolve variable length PROLOG programs for simple problems.

In 1989 Koza used Lisp S-Expressions to evolve programs in a process he called Genetic Programming [110]. Koza adopted a parse tree based representation for the individuals in the population. This allowed the representation of variable-sized individuals that took the form of actual computer programs. The genetic operators in this case, as with Cramer, are applied to sub-trees. GP is a widely adopted paradigm for the evolution of programs in EC today. Table 1.1 summarises these milestones. A more in-depth account of development of EC is given in [67].

From the very outset of the modern field of GP researchers incorporated the use of grammars to overcome the problem of closure. This is where the function set must be well defined in order to allow crossover between any two

Table 1.1 Chronological summary of major developments in EC up to GP.

Year	Researcher	Contribution
1948	Turing	Observation of paradigm
1958	Friedberg	Learning Machine, mutation
1966	Fogel, Owens & Walsh	Evolution of Finite State Machines
1975	Holland	Genetic Algorithm
1985	Cramer	GP with Trees and Linear Representations
1986	Hicklin, Fujiki	Grammars for GP
1987	Dickmanns & Schmidhuber	Evolution of PROLOG programs
1989	Koza	Evolution of Lisp S-Expressions (GP)

sub-trees as the function signatures must be syntactically the same. A second wave of grammars followed in the 1990s with Whigham [230] using a Context Free Grammar (CFG) to overcome closure by representing individuals as derivation trees. Wong and Leung [233, 232] follow a similar technique with the evolution of derivation trees in the spirit of Inductive Logic Programming adopting Logic Grammars. In parallel, Banzhaf [9] developed a linear form of GP, known as Binary GP (BGP), where genotype-phenotype mapping occurs from a binary string genome to a functional program phenotype. 1998 saw the first publication on Grammatical Evolution (GE) [190].

GE [148] draws inspiration from both molecular biology and the use of formal grammars. Like GAs and the work of Banzhaf, a linear genome is employed. In the case of GE, the search and solution spaces are decoupled and like GP, variable length individuals are used. This is achieved through the use of a genotype-phenotype mapping where the linear genome selects production rules from a Backus Naur Form (BNF) grammar to map down to a syntactically correct program. A more detailed description of GE is available in Chapter 2.

GE is a relatively new addition to the area of EC, and as such its utility in dynamic environments has not been fully explored. GE presents interesting avenues for exploration in dynamic environments over the more established algorithms in the field.

- The decoupling of the search and solution spaces provides a novel perspective in the monitoring of diversity within the population. Due to the genotype-phenotype mapping, many genotypes can map to a single phenotype giving two levels of diversity. Yet genetic parents that have the same phenotype can produce different offspring. The maintaining of a high level of diversity within the population is important in an adaptive system as it allows the potential for the discovery of new solutions when the target changes. This decoupling also leads to a ripple effect with the crossover genetic operator, which counteracts some of the issues of neutral crossover in a converged population while adding diversity in the phenotype space.
- The addition of a BNF grammar affords the modeler a very high degree of control in the evolutionary process. Discrete atomic units can be defined in

the grammar, minimising the destructive impact of some genetic operators. Conversely, a very granular grammar can be defined, presenting evolution with a widely scoped search space.
- In addition another dimension of adaptability can be explored by adapting the BNF grammar itself over the course of the evolutionary process. This allows evolution itself to incorporate biases into the solution space as it sees fit.

In this book we will explore and extend these avenues for the purposes of adapting in a dynamic environment.

1.3 Foundations

Considering how little research has been conducted into the use of Genetic Programming systems in dynamic environments the aim of this book is to begin to develop an understanding of how they, and more specifically GE, behave in these environments. In establishing this understanding certain questions must be answered focusing on three characteristics, adaptability, memory, and robustness:

- *Is the correct infrastructure in place for GE to navigate dynamic environments?* This question examines what practical extensions are required to GE in order for it to evolve a robust competent population for dynamic environments.
- *Is GE capable of discovering new solutions when change in the environment occurs?* The goal is to examine if GE converges to an optimal solution and fails to adapt or navigate to a new and better solution once a change has occurred.
- *Can GE maintain a diverse population of robust solutions capable of handling dynamic data?* Through being exposed to dynamic data GE should be able to produce robust generalised solutions capable of navigating new data or shifts in the fitness landscape as ideally it will have mined underlying information as it evolves across heterogeneous data.

Related to these are the wider EC issues of what role memory plays in dynamic environments. How is diversity linked to a shift or change in the environment? And what constitutes a good testbed for analysing evolutionary algorithms in a dynamic manner? We start to address these questions throughout this book and in so doing we make contributions across a range of domains including Grammatical Evolution, EC, and Finance.

For the first time we analyse GE in a dynamic environment. Adapting a population as a target changes over time opens up issues separate to those encountered by conducting evolution over a static data set. This book aims to close the knowledge gaps in the domain of GE. As already asserted, evolution is a process which has produced a rich and diverse range of biological organisms in the ever-changing natural world. In a step to realising the full potential of GE, this book places GE in that setting and produces recommendations based on experimental evidence.

Evolvability concerns the ease with which new fitter solutions may be evolved. In dynamic environments the provision of a high level of evolvability is key in enabling the population to track a moving optimum. GE's structure, where it incorporates a BNF grammar and adopts a genotype-to-phenotype mapping, presents two levels of evolvability. One where the neutral mutations may occur over the many-to-one mapping brought about from the genotype to the phenotype spaces enabling the evolutionary search to continue while maintaining phenotypic fitness. The other at the BNF grammar level attention can be paid to designing production rules that facilitates incremental improvements in fitness as the symbols are adapted, in essence providing a smooth hill climbing mechanism.

1.4 Structure of This Book

The remainder of the book is structured as follows. An introduction to Grammatical Evolution is presented in Chapter 2. Chapter 3 contains a comprehensive survey of literature relevant to dynamic environments in EC. The various threads of innovation are constructively grouped together and a critical analysis is developed for each. Resulting from this analysis, a unifying taxonomy of the different works is presented. Given this broader context in evolutionary computation, in subsequent chapters we build upon these foundations to begin our analysis and the extension of GE to dynamic environments. Chapter 4 provides an initial discussion on this topic.

Considering virtually all "solutions" contain constants or mathematical expressions, the creation and adaptation of constants in any setting is of immense importance. When the setting is that of a dynamic environment, then even more emphasis is placed on the ability to create new constants and adapt existing constants. An area that has not received much attention for GE specifically is how best to create and adapt constants in the system. By utilising the properties and flexibility of BNF grammars, which are unique to GE, we describe the best method for constant creation and adaptation in GE, and novel methods are developed and explored in Chapters 5 and 6. The first new method of constant creation, which is superior to the standard method adopted in GP (ERCs), is called Persistent Random Constants. The second is capable of continuous constant creation along with the adaptation of individual constants and the formation of expressions, is called Digit Concatenation. Finally, a meta-grammar approach is also analysed.

Chapters 7 and 8 present the application of Grammatical Evolution to financial trading. In the first instance we examine behaviour in a static environment to obtain a baseline in terms of performance and behaviour of the algorithm, before moving on to its application to a dynamic version of the problem. Of importance to evolution in dynamic environments is that an adequate level of diversity is maintained. A wide spread across the solution space allows a population to converge on new areas when the optimum changes. We demonstrate that the population of solutions naturally increases its diversity with a shift in the optimum. If a population is to be able to adapt to a new change, then it is important that the representation be conducive to producing new individuals with potentially new perspectives, as an environment changes.

Our research indicates that where a population of solutions are evolved over static data, brittle solutions are produced. Where this data is dynamic the population is exposed and evolved over heterogeneous data providing scope to mine underlying persistent trend information without overfitting to specific features that may arise on a single set of data alone. Experiments conducted show that analysis of historical data provides useful information on how to act on current data. Historical time series for financial markets are analysed with the system at hand outperforming random trading strategy and suggesting these markets are not strongly efficient. This finding has implications for the use of technical analysis and supports the strategy of in depth research of historical data in order to identify *alpha* (an edge) in financial markets.

Finally we end the book with some Conclusions and point out some directions for Future research in this area in Chapter 9.

Chapter 2
Grammatical Evolution

Genetic Programming (GP) is a population-based search method based upon neo-Darwinian principles of evolution, which manipulates executable structures. It can be used to automatically generate computer code to solve real-world problems in a broad array of application domains [173].

Grammatical Evolution(GE) (see [154, 148, 24, 54, 55, 151, 190, 160, 86, 25, 43, 7, 163]) is a grammar-based form of GP. It marries principles from molecular biology to the representational power of formal grammars. GE's rich modularity gives a unique flexibility, making it possible to use alternative search strategies, whether evolutionary, or some other heuristic (be it stochastic or deterministic) and to radically change its behaviour by merely changing the grammar supplied. As a grammar is used to describe the structures that are generated by GE, it is trivial to modify the output structures by simply editing the plain text grammar. The explicit grammar allows GE to easily generate solutions in any language (or a useful subset of a language). For example, GE has been used to generate solutions in multiple languages including Lisp, Scheme, C/C++, Java, Prolog, Postscript, and English. The ease with which a user can manipulate the output structures by simply writing or modifying a grammar in a text file provides an attractive flexibility and ease of application not as readily enjoyed with the standard approach to Genetic Programming. The grammar also implicitly provides a mechanism by which type information can be encoded thus overcoming the property of closure, which limits the traditional representation adopted by Genetic Programming to a single type. The genotype-phenotype mapping also means that instead of operating exclusively on solution trees, as in standard GP, GE allows search operators to be performed on the genotype (e.g., integer or binary chromosomes), in addition to partially derived phenotypes, and the fully formed phenotypic derivation trees themselves. As such, standard GP tree-based operators of subtree-crossover and subtree-mutation can be easily adopted with GE.

In this chapter an introduction to GE is given including a description of the genotype-to-phenotype mapping process in Section 2.2, which is a key

I. Dempsey et al.: Foundations in Gramm. Evol. for Dyn. Environ., SCI 194, pp. 9–24.
springerlink.com

feature of GE in dealing with dynamic environments. Section 2.3 describes how the mutation and crossover genetic operators are applied in GE, and is followed by an introduction to the adoption of alternative search engines in Section 2.4. Section 2.5 details various applications of GE. The chapter then closes with some brief conclusions.

2.1 An Overview of Grammatical Evolution

GE is a grammar-based Genetic Programming paradigm that is capable of evolving programs or rules in any language [148, 151, 154, 190]. GE is similar to Genetic Programming as popularised by Koza [110], in that it uses an evolutionary process to automatically generate variable-length computer programs. Unlike GP however, GE adopts a population of linear genotypic binary or integer strings, which are transformed into functional phenotypic programs through a genotype-to-phenotype mapping process [9]. This transformation is governed through the use of a BNF grammar, which specifies the language of the produced solutions. The mechanics of the mapping process are discussed in more detail in Section 2.2.

The mapping process creates a distinction between the search space and the solution space. Genotype strings are evolved with no knowledge (or respect) of their phenotypic equivalent, apart from a fitness measure, in an unconstrained evolutionary search. These genotypic individuals are then projected into the constrained phenotypic solution space through the use of the mapping process. In undergoing evolution and reproduction, the canonical genetic operators of Holland's GA may be applied to the genotype strings. The application of these operators has been shown to create a phenotypic "ripple effect". Evidence suggests that this effect promotes a useful exchange of derivation sub-sequences during crossover events [154], while by the same measure contributing to phenotypic diversity.

GE is modular in design, where the grammar, the search engine and the objective function all represent plug-in components to GE. Figure 2.1 presents the modular design of GE. Such modularity allows it to be used in conjunction with researchers' prefered search algorithms. In recent developments GE has been combined with a swarm algorithm [163] and a differential evolution algorithm [162]. In other developments, the potential of the grammar component has been explored with meta-Grammars [156] where a universal grammar, or grammar's grammar is defined, allowing the evolution of a solution's own vocabulary. meta-Grammars have also been used in the development of the meta-Grammar GA (mGGA) [159] that encourages the evolution of building blocks (see Section 3.3.5) which can be reused to construct solutions more efficiently. More detail on meta-Grammars combined with GE is given in Chapter 6. Also on research into grammars, attribute grammars have been examined [42, 158] that enable the addition of semantic and context-sensitive information into the grammars.

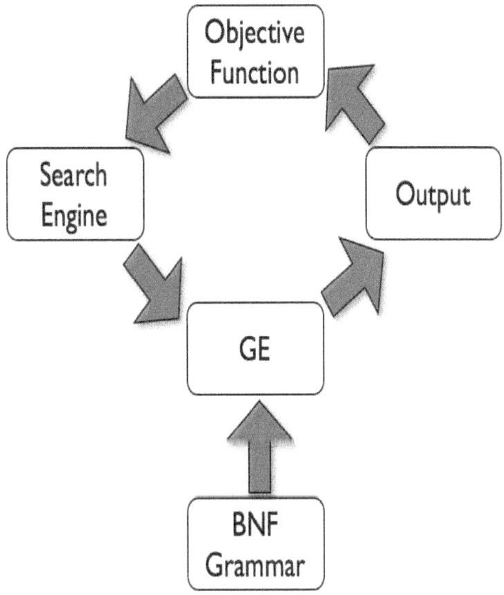

Fig. 2.1 GE's modular design and workflow

2.2 Mapping Process

When approaching a problem using GE, initially a BNF grammar must be defined. This grammar specifies the syntax of desired phenotypic programs to be produced by GE. The syntax at its broadest may detail the specification of a programming language such as C, or more practically a subset of such a language. The development of a BNF grammar also affords the researcher the ability to incorporate domain biases or domain-specific functions.

A BNF grammar is made up of the tuple N, T, P, S; where N is the set of all non-terminal symbols, T is the set of terminals, P is the set of production rules that map N to T, and S is the initial start symbol and a member of N. Where there are a number of production rules that can be applied to a non-terminal, a " | " (or) symbol separates the options. Using

```
N = {<expr>, <op>, <operand>, <var>}
T = {1, 2, 3, 4, +, -, /, *, x, y}
S = {<expr>}
```

with P, below is an example of a BNF grammar:

```
<expr>::=  <expr><op><expr>         (0)
         | <operand>                (1)
<op>::=  +                          (0)
       | -                          (1)
       | /                          (2)
       | *                          (3)
<operand>::=  1                     (0)
```

```
             | 2                      (1)
             | 3                      (2)
             | 4                      (3)
             | <var>                  (4)
<var>::=  x                           (0)
        | y                           (1)
```

Using such a grammar as an input, GE then employs the expression

$$Rule = c\%r$$

that selects an option from the production rule for the symbol being currently mapped; where c is the codon value and r is the number of production rules available for the current non-terminal. An example of the mapping process employed by GE is shown in Figure 2.2. Starting with a binary string, an integer string is evolved, typically by using 8 bits per codon, which is a sequence of genes. These are then used to choose rules from a given BNF grammar and to generate a phenotypic program.

Beginning with a given *start symbol* `<E>`, a production associated with that symbol is chosen to replace it, by using the current codon from the integer string. In the example, the codon 8 is mapped to the number of available productions associated with `<E>`, of which there are 6. So therefore $8\ mod\ 6 = 2$, and `<E>` is replaced with the sequence `(- <E><E>)`.

The next step consists of choosing a production for the next `<E>` non-terminal symbol, which is now the leftmost symbol on the phenotype string under construction. This is done by using the next codon, 4, and again mapping it to the number of productions associated with `<E>`, giving $4\ mod\ 6 = 4$, so `<E>` is replaced with `(+ <E><E>)`.

The mapping process continues in this manner, always replacing the leftmost non-terminal symbol with a production associated with that symbol on the grammar, chosen by a codon. The mapping in standard GE terminates when one of the following conditions is met:

1. A complete program is generated before the end of the genome is encountered. This occurs when all the non-terminals in the expression being mapped are transformed into elements from the terminal set of the BNF grammar as seen in the example in Figure 2.2.
2. The end of the genome is reached, in which case the *wrapping* operator [155] is invoked. This results in the return of the genome-reading frame to the left-hand side of the genome once again. The reading of codons will then continue as before unless an upper threshold, representing the maximum number of wrapping events, is reached during the mapping process.
3. A threshold on the number of wrapping events is reached and the individual is still not completely mapped. In this case the mapping process is halted and the individual is assigned the worst possible fitness value.

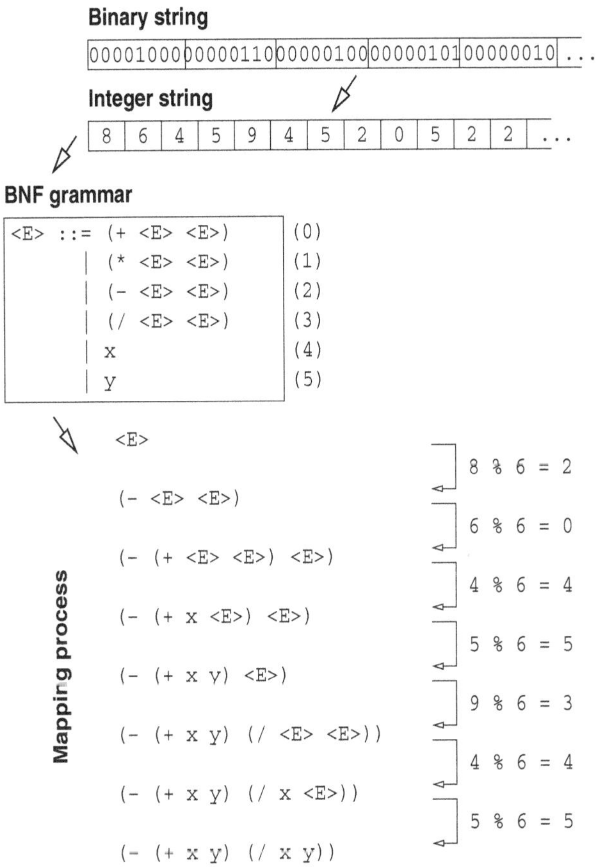

Fig. 2.2 The GE mapping process. A binary string is converted into an integer string, using 8 bits per codon. These integers are then used to choose productions from a BNF grammar, which map a given start symbol to a sequence of terminal symbols.

A variation on the termination of the mapping process is provided in Genr8 [87] where, upon reaching the end of the genotype wrapping occurs, but only production rules that result in terminating sequences are allowed to be selected. Effectively the grammar is modified to remove rules that expand the developing sentence.

2.2.1 πGE

A recent innovation in GE's mapping process has led to the development of Position Independent GE or πGE [157]. In standard GE, there is a positional dependency as the mapping process moves from left to right in consuming the non-terminals. πGE removes this dependency whereby instead of representing each codon in standard GE as a single value, codons in πGE have two values: *nont* and *rule*. In this case, *nont* contains the encoding to select which non-terminal is to be consumed by the mapper. It does this using a rule similar to the original mapping rule:

$$NT = nont\%count$$

where NT is the non-terminal to be consumed (counted 0,1,...,n from left to right of the remaining non-terminals), *nont* is the value from the individual, and *count* is the number of non-terminals remaining. The rule part of the codon pair then, as in the previous section, selects which production rule to apply to the chosen non-terminal. For example, given that the grammar in the previous section produces the sequence `<exp><op><exp>` after the first symbol is consumed, these non-terminals would be counted `<exp>` – 0, `<op>` – 1 and `<exp>` – 2, giving 3 non-terminals. If the *nont* value was 8, this would yield an expression 8%3 selecting non-terminal 2 as the next to be consumed.

This more open-ended form of the mapping process was found to result in impressive performance gains on a number of benchmark problems.

2.3 Mutation and Crossover in GE

As already described, the representation of the individuals to which the genetic operators in GE are applied are variable-length linear strings. Due to this representation, the mechanics of the canonical GA operators are the same; mutation changes a bit or an integer to another random value, and one-point crossover swaps sections of the genetic code between parents. However, because of the mapping process, the effect on the phenotype can be complex.

When mutation is applied in standard GP, it occurs at the phenotypic level where the selected node or sub-tree is regenerated randomly and obeys the closure requirements of GP. In GE and Holland's GA, this operator is applied without constraints. However, in GE the potential exists for the mutation to have no effect, or to be *neutral*, at the phenotypic level [186]. For example, given the following BNF production rule:

```
<var>::= x                    //(0)
       | y                    //(1)
```

where the non-terminal `<var>` can be replaced with the variables `x` or `y`, and an integer codon value of 20. When `<var>` is consumed, it will be replaced with the variable `x` because, using the mapping rule from the previous section, the rule to be executed is 20 % 2; resulting in 0 or the first production rule. If a mutation occurs and changes the codon to 21, `<var>` will be replaced with `y`. However, if this codon is changed to 22, or any other even number for that matter, (as there are just two production rules) the phenotype will remain as x. This means that neutral mutation is occurring as the resulting functionality of the expressed gene remains the same.

One-point crossover in GE is conducted under the same mechanics as its GA counterpart, where the genetic material to the right of the selected crossover points is swapped between two parents. In the case of GE, again due to the mapping process, the effect at the phenotypic level is different from that of GAs. Crossover in GE has a ripple effect on the derivation sequence after the crossover point. Figure 2.3 helps to illustrate this process. The mapping process in GE can be described as a sequence of derivations, which in turn can be represented as a derivation tree. At the core of crossover in GE

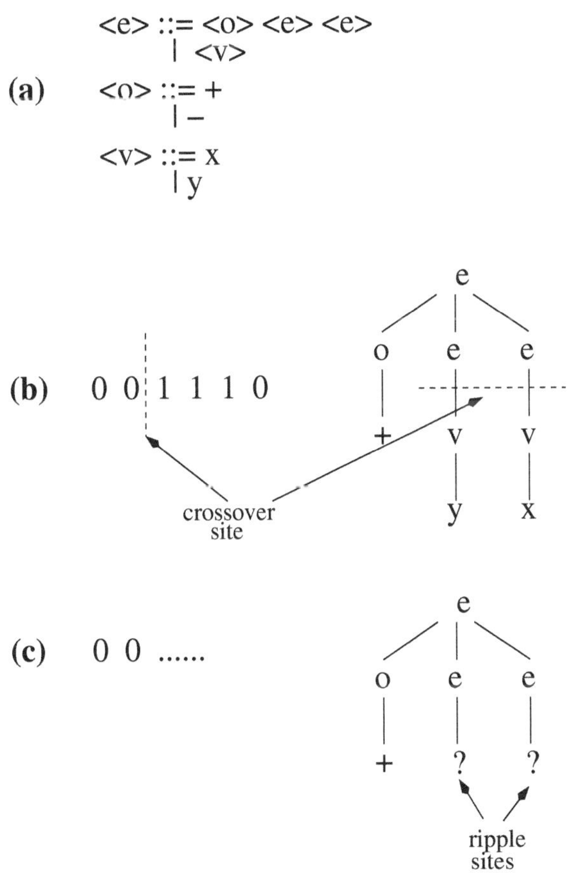

Fig. 2.3 Illustration of derivation tree during crossover. The BNF grammar (a) is applied to the genotype in (b) to produce the derivation tree on the right. A crossover site is selected after the first two bits with the resulting effect observed on the derivation tree in (c). The genotypic sequence which is added after the crossover point will be placed in a context at the ripple sites (denoted by the "?"s) which may differ from its original context.

is that when genetic material is swapped to another individual, the context in which it is placed may be different. This results in a different phenotypic expression compared to when it was in its original placement. While this form of crossover can appear to be destructive, research has shown that useful derivation sequences are transfered [155]. When analysing the behaviour of this *ripple crossover* it is clear that it exchanges multiple subtrees, and consequently significantly more material between solutions than standard sub-tree crossover, with 50% of an individual being exchanged on average. Further work by Harper & Blair [85] has also led to the development of a structure preserving two-point crossover that extracts information from the BNF grammar to minimise the destructive impact to useful blocks of code on either side of the crossover sub-section. This form of crossover is similar

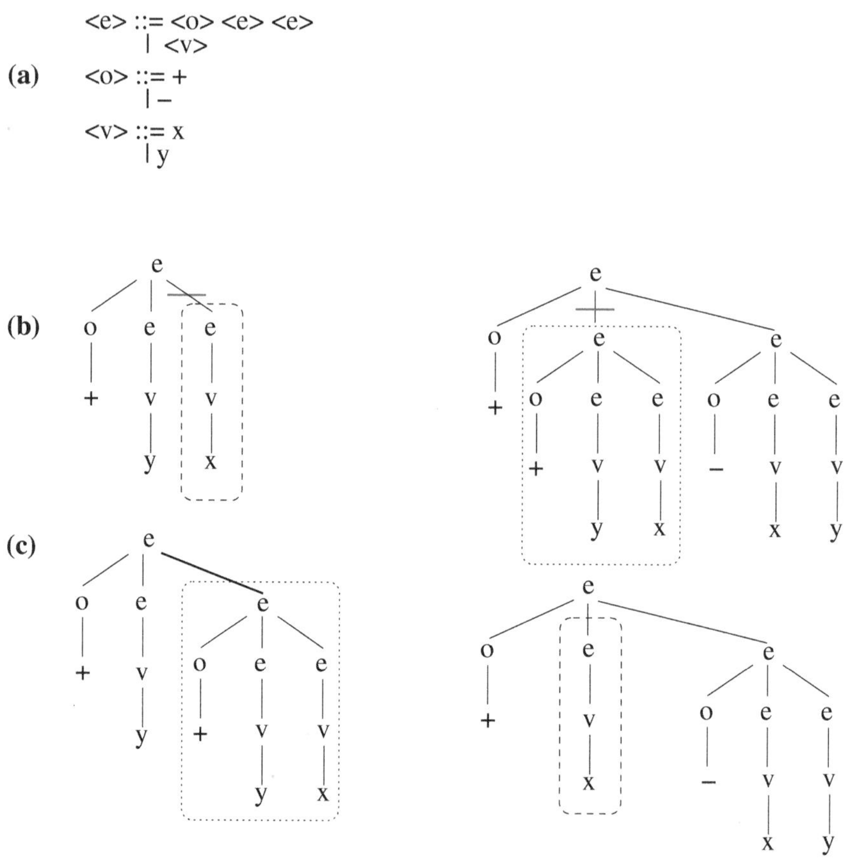

Fig. 2.4 Illustration of derivation tree during subtree crossover in GE. The BNF grammar (a) is applied to the genotype in (b) to produce the derivation tree on the right. A crossover site is selected after the first two bits with the resulting effect observed on the derivation tree in (c).

to sub-tree crossover in tree-based GP and an example of its operation illustrated in Figure 2.4.

Armed with more in-depth knowledge on the impact of GE's search operators to phenotypic change it is possible to design an appropriately balanced set of operators. Like any other form of EC it is essential that the genetic operators allow provision for the generation of both small and large changes, and to balance exploitation and exploration in the process. This is an active area of research within GE.

2.4 Alternative Search Engines

We hinted in the introduction to this chapter that it is possible to use alternative search strategies to the variable-length Evolutionary Algorithm of GE. Two prominent examples of this are evidenced by the use of Particle Swarm Optimisation and Differential Evolution to create the Grammatical Swarm (GS) [163] and Grammatical Differential Evolution (GDE) [162] variants. Note that we modify the second half of the algorithm's name to denote the search engine employed. We now provide a brief introduction to GS and GDE.

2.4.1 Grammatical Swarm

One model of social learning that has attracted interest in recent years is drawn from a swarm metaphor. Two popular variants of swarm models exist, those inspired by studies of social insects such as ant colonies, and those inspired by studies of the flocking behavior of birds and fish. The essence of these systems is that they exhibit flexibility, robustness and self-organisation [14]. Although the systems can exhibit remarkable coordination of activities between individuals, this coordination does not stem from a 'center of control' or a 'directed' intelligence, rather it is self-organising and emergent. 'Social Swarm' researchers have emphasized the role of social learning processes in these models [105, 106]. In essence, social behavior helps individuals to adapt to their environment, as it ensures that they obtain access to more information than that captured by their own senses.

In the context of PSO, a swarm can be defined as '... a population of interacting elements that is able to optimize some global objective through collaborative search of a space.' [105](p. xxvii). The nature of the interacting elements (particles) depends on the problem domain, in this study they represent program construction rules. These particles move (fly) in an n-dimensional search space, in an attempt to uncover ever-better solutions to the problem of interest.

Each of the particles has two associated properties, a current position and a velocity. Each particle has a memory of the best location in the search space that it has found so far (p_{best}), and knows the best location found to date by all the particles in the population (g_{best})(or in an alternative version of the algorithm, a neighbourhood around each particle). At each step of the algorithm, particles are displaced from their current position by applying a velocity vector to them. The velocity size / direction is influenced by the velocity in the previous iteration of the algorithm (simulates 'momentum'), and the location of a particle relative to its p_{best} and g_{best}. Therefore, at each step, the size and direction of each particle's move is a function of its own history (experience), and the social influence of its peer group.

A number of variants of the particle swarm algorithm (PSA) exist. The following paragraphs provide a description of a basic *continuous* version of the algorithm.

i. Initialise each particle in the population by randomly selecting values for its location and velocity vectors.
ii. Calculate the fitness value of each particle. If the current fitness value for a particle is greater than the best fitness value found for the particle so far, then revise p_{best}.
iii. Determine the location of the particle with the highest fitness and revise g_{best} if necessary.
iv. For each particle, calculate its velocity according to equation 2.1.
v. Update the location of each particle according to equation 2.3.
vi. Repeat steps ii - v until stopping criteria are met.

The update algorithm for particle i's velocity vector v_i is:

$$v_i(t+1) = (w * v_i(t)) + (c_1 * R_1 * (p_{best} - x_i)) + (c_2 * R_2 * (g_{best} - x_i)) \quad (2.1)$$

where,

$$w = wmax - ((wmax - wmin)/itermax) * iter \quad (2.2)$$

In equation 2.1, p_{best} is the location of the best solution found to-date by particle i, g_{best} is the location of the global-best solution found by all particles to date, c_1 and c_2 are the weights associated with the p_{best} and the g_{best} terms in the velocity update equation, x_i is particle i's current location, and R_1 and R_2 are randomly drawn from U(0,1). The term w represents a momentum coefficient which is reduced according to equation 2.2 as the algorithm iterates. In equation 2.2, *itermax* and *iter* are the total number of iterations the algorithm will run for, and the current iteration value respectively; *wmax* and *wmin* set the upper and lower boundaries on the value of the momentum coefficient. The velocity update on any dimension is constrained to a maximum value of *vmax*. Once the velocity update for particle i is determined, its position is updated (equation 2.3, and p_{best} is updated if necessary (equations 2.4 & 2.5).

$$x_i(t+1) = x_i(t) + v_i(t+1) \quad (2.3)$$

$$y_i(t+1) = y_i(t) \text{ if, } f(x_i(t)) \leq f(y_i(t)) \quad (2.4)$$

$$y_i(t+1) = x_i(t) \text{ if, } f(x_i(t)) > f(y_i(t)) \tag{2.5}$$

After the location of all particles have been updated, a check is made to determine whether g_{best} needs to be updated (equation 2.6).

$$\hat{y} \in (y_0, y_1, ..., y_n) | f(\hat{y}) = \max\ (f(y_0), f(y_1), ..., f(y_n)) \tag{2.6}$$

In Grammatical Swarm (GS) the update equations for the swarm algorithm are as described earlier, with additional constraints placed on the velocity and particle location dimension values, such that maximum velocities $vmax$ are bound to ± 255, and each dimension is bound to the range [0,255] (denoted as $cmin$ and $cmax$ respectively). Note that this is a continuous swarm algorithm with real-valued particle vectors. The standard GE mapping function is adopted, with the real-values in the particle vectors being rounded up or down to the nearest integer value for the mapping process. In the current implementation of GS, fixed-length vectors are adopted, within which it is possible for a variable number of dimensions to be used during the program construction genotype-phenotype mapping process. A vector's elements (values) may be used more than once if wrapping occurs, and it is also possible that not all dimensions will be used during the mapping process if a complete program comprised only of terminal symbols, is generated before reaching the end of the vector. In this latter case, the extra dimension values are simply ignored and considered introns that may be switched on in subsequent iterations.

GS Experimental Findings

A diverse selection of benchmark programs from the literature were tackled to demonstrate proof of concept for the GS method. The problems included Santa Fe Ant Trail, a Symbolic Regression instance ($x + x^2 + x^3 + x^4$), the 3-Multiplexer boolean problem, and Mastermind. The parameters adopted across the experiments were $c_1 = c_2 = 1.0$, $wmax = 0.9$, $wmin = 0.4$, $cmin = 0$ (minimum value a coordinate may take), $cmax = 255$ (maximum value a coordinate may take). In addition, a swarm size of 30 running for 1000 iterations using 100 dimensions is used. The same problems are also tackled with GE in order to determine how well GS is performing at program generation in relation to the more traditional variable-length Genetic Algorithm search engine of standard GE. In an attempt to achieve a relatively fair comparison of results given the differences between the search engines of Grammatical Swarm and Grammatical Evolution, we have restricted each algorithm in the number of individuals they process. Grammatical Swarm running for 1000 iterations with a swarm size of 30 processes 30,000 individuals, therefore, a standard population size of 500 running for 60 generations is adopted for Grammatical Evolution. The remaining parameters for Grammatical

Evolution are roulette selection, steady state replacement, one-point crossover with probability of 0.9, and a bit mutation with probability of 0.01.

Table 2.1 provides a summary and comparison of the performance of GS and GE on each of the problem domains tackled. 100 independent runs were performed for the data reported. In two out of the four problems GE outperforms GS, and GS outperforms GE on the other two problem instances. The key finding is that the results demonstrate proof of concept that GS can successfully generate solutions to problems of interest. In this initial study, we have not attempted parameter optimisation for either algorithm, but results and observations of the particle swarm engine suggests that swarm diversity is open to improvement. We note that a number of strategies have been suggested in the swarm literature to improve diversity [202], and we suspect that a significant improvement in GS performance can be obtained with the adoption of these measures. Given the relative simplicity of the Swarm algorithm, the small population sizes involved, and the complete absence of a crossover operator synonymous with program evolution in GP, it is impressive that solutions to each of the benchmark problems have been obtained.

Table 2.1 A comparison of the results obtained for Grammatical Swarm and Grammatical Evolution across all the problems analysed

	Mean Best Fitness (Std.Dev.)	Mean Average Fitness (Std.Dev.)	Successful Runs
Santa Fe ant			
GS	75.24 (16.64)	33.43 (3.69)	43
GE	80.18 (13.79)	46.43 (11.18)	58
Multiplexer			
GS	0.97 (0.05)	0.87 (0.01)	79
GE	0.95 (0.06)	0.88 (0.04)	56
Symbolic Regression			
GS	0.31 (0.35)	0.07 (0.02)	20
GE	0.88 (0.30)	0.28 (0.28)	85
Mastermind			
GS	0.91 (0.04)	0.88 (0.01)	18
GE	0.90 (0.03)	0.89 (0.00)	10

2.4.2 Grammatical Differential Evolution

Differential evolution (DE) [209, 210, 211, 175] is a population-based search algorithm. The algorithm draws inspiration from the field of Evolutionary Computation, as it embeds implicit concepts of mutation, recombination and fitness-based selection, to evolve from an initial randomly generated population to a solution to a problem of interest. It also borrows principles from

Social Algorithms through the manner in which new individuals are generated. Unlike the binary chromosomes typical of GAs, an individual in DE is generally comprised of a real-valued chromosome.

Although several DE algorithms exist we only describe one version of the algorithm based on the *DE/rand/1/bin* scheme [209]. The different variants of the DE algorithm are described using the shorthand DE/x/y/z, where x specifies how the base vector to be perturbed is chosen (*rand* if it is randomly selected or *best* if the best individual is selected), y is the number of difference vectors used, and z denotes the crossover scheme used (*bin* for crossover based on independent bi-nominal experiments, and *exp* for exponential crossover).

At the start of this algorithm, a population of N, d-dimensional vectors $X_j = (x_{i1}, x_{i2}, \ldots, x_{id})$, $j = 1, \ldots, n$, is randomly initialised and evaluated using a fitness function f. During the search process, each individual (j) is iteratively refined. The modification process has three steps:

i. Create a variant solution, using randomly selected members of the population.
ii. Create a trial solution, by combining the variant solution with j (crossover step).
iii. Perform a selection process to determine whether the trial solution replaces j in the population.

Under the mutation operator, for each vector $X_j(t)$, a variant solution $V_j(t+1)$ is obtained using equation 2.7:

$$V_j(t+1) = X_m(t) + F(X_k(t) - X_l(t)) \tag{2.7}$$

where $k, l, m \in 1, \ldots, N$ are mutually different, randomly selected indices, and all the indices $\neq j$ (X_m is referred to as the base vector, and $X_k(t) - X_l(t)$ is referred to as a difference vector). Variants on this step include the use of more than three individuals from the population, and/or the inclusion of the highest-fitness point in the population as one of these individuals [209]. The difference between vectors X_k and X_l is multiplied by a scaling parameter F (typically, $F \in (0, 2]$). The scaling factor controls the amplification of the difference between X_k and X_l, and is used to avoid stagnation of the search process. Following the creation of the variant solution, a trial solution $U_j(t+1) = (u_{j1}, u_{j2}, \ldots, u_{jd})$ is obtained from equation 2.8.

$$U_{jn}(t+1) = \begin{cases} V_{jn}, \text{if } (rand \leq CR) \text{ or } (j = rnbr(i)) ; \\ X_{jn}, \text{if } (rand > CR) \text{ and } (j \neq rnbr(i)). \end{cases} \tag{2.8}$$

where $n = 1, 2, \ldots, d$, *rand* is drawn from a uniform random number generator in the range (0,1), CR is the user-specified crossover constant from the range (0,1), and $rnbr(i)$ is a randomly chosen index chosen from the range $(1, 2, \ldots, n)$. The random index is used to ensure that the trial solution differs by at least one component from $X_i(t)$. The resulting trial solution replaces its

Fig. 2.5 A representation of the Differential Evolution variety-generation process. The value of F is set at 0.50. In a simple 2-d case, the child of particle X_j can end up in any of three positions. It may end up at either of the two positions X_j^*, or at the position of particle $V_j(t+1)$.

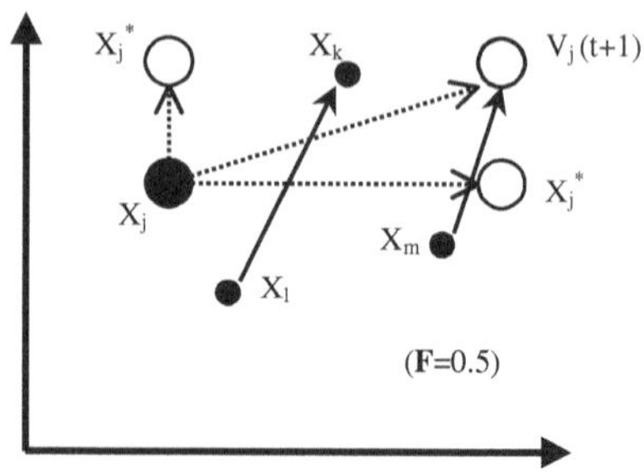

predecessor , if it has higher fitness (a form of selection), otherwise the predecessor survives unchanged into the next iteration of the algorithm (equation 2.9).

$$X_i(t+1) = \begin{cases} U_i(t+1), & \text{if } f(U_i(t+1)) < f(X_i(t)); \\ X_i(t), & \text{otherwise.} \end{cases} \tag{2.9}$$

Fig. 2.5 provides a graphic of the adaptive process of GDE. The DE algorithm has three parameters, the population size (N), the crossover rate (CR), and the scaling factor (F). Higher values of CR tend to produce faster convergence of the population of solutions. Typical values for these parameters are in the range, N=50-100 (or ten times the number of dimensions in a solution vector), CR=0.8-0.9 and F=0.3-0.5.

Grammatical Differential Evolution (GDE) adopts a Differential Evolution learning algorithm coupled to a Grammatical Evolution (GE) genotype-phenotype mapping to generate programs in an arbitrary language. The standard GE mapping function is adopted with the real-values in the vectors being rounded up or down to the nearest integer value, for the mapping process. In the current implementation of GDE, fixed-length vectors are adopted within which it is possible for a variable number of elements to be required during the program construction genotype-phenotype mapping process. A vector's values may be used more than once if the wrapping operator is used, and in the opposite case it is possible that not all elements will be used during the mapping process if a complete program comprised only of terminal symbols is generated before reaching the end of the vector. In this latter case, the extra element values are simply ignored and considered introns that may be switched on in subsequent iterations.

GDE Experimental Findings

The same diverse set of problems are tackled with GDE as with GS, including an instance of Symbolic Regression ($x+x^2+x^3+x^4$), the Santa Fe Ant Trail, boolean 3-Multiplexer, and Mastermind. The parameters adopted across the following experiments are Params of GDE....popsize 500, 100 iterations, strlen 100, F=0.9, CR=1.0, DE/best/1/exp. Gene values are bound to the range $[0 \rightarrow 255]$.

The same problems are also tackled with Grammatical Evolution in order to get some indication of how well GDE is performing at program generation in relation to the more traditional variable-length Genetic Algorithm-driven search engine of standard GE. a standard population size of 500 running for 60 generations is adopted for Grammatical Evolution. The remaining parameters for Grammatical Evolution are roulette selection, steady state replacement, one-point crossover with probability of 0.9, and a bit mutation with probability of 0.01.

Table 2.2 provides a summary and comparison of the performance of Grammatical Differential Evolution, and Grammatical Evolution on each of the problem domains tackled. The reported results are averaged over 50 independent runs. In three out of the four problems Grammatical Evolution outperforms GDE. The key finding is that the results demonstrate proof of concept that GDE can successfully generate solutions to problems of interest.

Table 2.2 A comparison of the results obtained for Grammatical Differential Evolution and Grammatical Evolution across all the problems analysed

	Santa Fe Ant	**Multiplexer**	**Symbolic Regression**	**Mastermind**
GDE/rand/1/bin	10	23	6	0
GDE/best/1/exp	7	**27**	4	0
GDE/rand-to-best/1/exp	9	**27**	4	0
GDE/rand-to-best/1/bin	7	25	5	0
GE	**17**	15	**24**	**3**

2.5 Applications of GE

Since its inception GE has received considerable attention and been applied to a wide variety of problem domains. Early studies saw GE being successfully applied to symbolic-regression problems [190], the evolution of trigonometric identities [192], the evolution of caching algorithms [145], and behavioural robotics [146, 147].

GE was extended and applied to the domain of surface design in [87] where it was combined with GENR8, a surface design tool that uses an evolutionary process to adapt surfaces to meet a user's specification. In the field of Bioinformatics, GE was applied to recognise eukaryotic promotors [161] that help in the identification of biological genes. While in [137] GE was used to evolve neural networks for feature selection in genetic epidemiology.

In the area of sound synthesis and analysis Ortega et al. [168] use GE to automatically generate compositions which were comparable to human composed works. In [109] GE was tasked with evolving phonological rules which can be used to translate text into graphemes and recognition of sounds as words.

Fractal curves of a high dimensionality were evolved in [169] and in [130] Petri-Net models of complex systems were evolved. GE has also been used in the design Logos [164].

In the financial domain, GE has been applied in a number of areas [24], with studies in foreign-exchange trading [17, 18, 22], bankruptcy and corporate analysis [16, 19, 20], and credit classification [21, 23]. Specific to the trading of market indices, which is examined in this book, studies have also been conducted [149, 150, 15] where the problem is examined in a static manner.

2.6 Conclusion

This chapter introduced Grammatical Evolution and described its mapping process and how the genetic operators of mutation and crossover are implemented along with their effects. Some noteable applications of GE were outlined.

In addition to the various papers on the subject (see the GP Bibliography [115]), further information on GE can be found from `http://www.grammatical-evolution.org` including links to various software implementations of GE. A version of GE in Java, GEVA [76], has recently been released by our Natural Computing Research & Applications group at University College Dublin. This is available directly from `http://ncra.ucd.ie/geva/`.

Subsequent chapters will explore the features of GE and its potential for use in dynamic environments. Before we begin a more in-depth analysis of GE in dynamic environments, in the following chapter we first provide an overview of the research to date on Evolutionary Computation in these non-stationary domains.

Chapter 3
Survey of EC in Dynamic Environments

3.1 Introduction

The domain of EC in dynamic environments can be broken up into four main areas. This chapter is therefore divided up into four sections so as to comprehensively detail the prior art in the field as it stands. Section 3.2 focuses on the definition of dynamic environments. It examines the different types of changes that can occur as well as the features of each type of change that differentiate it from the other. Section 3.3 identifies the various approaches researchers have adopted in attempting to tune the EC paradigm to dynamic environments. Following from this, Section 3.4 describes the difficulties encountered when trying to measure the performance of an evolutionary algorithm placed in a dynamic environment. It also covers the types of metrics adopted to date. Section 3.5 looks at the different benchmark problems explored in the literature. The chapter then continues with a review of the findings of this survey in Section 3.6.

A number of conclusions emerge from this chapter. It is found that while most of the research effort on EC in dynamic environments has been aimed at adding memory, this comes at a computational cost with mixed results. At the same time, it overlooks the idea that problem-solving competency can be built into the solution population in a general manner. The chapter finds that efforts to maintain a diverse or dispersed population are of significant importance for dynamic environments. However, of lesser importance is the field's emphasis on developing benchmark problems and metrics. Too many have been developed and all use bounded search spaces aimed at traditional GA fixed-length individuals while often also having a known optimum. These points are elaborated upon in Section 3.7.

3.2 Dynamic Problems

3.2.1 Background

A dynamic problem can most broadly be defined as a problem in which some element under its domain varies with the progression of time. Dynamic

I. Dempsey et al.: Foundations in Gramm. Evol. for Dyn. Environ., SCI 194, pp. 25–54.
springerlink.com

problems have received a growing degree of attention in recent years as authors have highlighted the fact that dynamic problems share more in common with real-world problems than static problems [78, 81, 27, 13]. Even though Goldberg first highlighted this link in 1987 [78], the momentum of research in the area only picked up towards the end of the 1990s. To date there are in the region of 200 publications available in the field of dynamic optimisation problems [59] compared to many thousands of publications in the area of EC as a whole [75].

This section will continue by surveying the relevant literature and categorising the types of problems that can occur in dealing with dynamic environments. Following from this a more detailed insight will be developed by incorporating the types of changes that can occur into a unifying table. Having analysed the characteristics of dynamic problems, the section then closes with an analysis of the differences between dynamic and static problems.

3.2.2 Dynamic Environment Categorisations

Taxonomies are useful categorisations that provide a conceptual framework for discussion, analysis, or information retrieval. A taxonomy should be discriminatory by allowing the separation of a group into mutually exclusive subgroups. This section examines two taxonomies relevant to EC in dynamic environments: types of dynamic problems and types of change.

Categories of Dynamic Problems

Trojanowski and Michalewicz (T&M), in their works [215, 216], develop a clear taxonomy of the range of problems to which EC is applied, including both static and dynamic. By splitting the problem in two and placing the solution landscape in one column and the constraints of the problem in the second, they discriminate between six categories of problems. Under the solution landscape or objective function, two possible states can be represented: either static or variable/dynamic. The constraints of a problem can then take on one of three states. They are non-existent or null, static, or variable. Table 3.1 lays out the various permutations.

The first two categories of problem in Table 3.1, represent the classes of problems to which the majority of EC research has been focused, where the objective function remains static over the lifetime of the problem. Problem categories 3–6 on the other hand, describe problems where some element in the problem domain changes, and form the area of study for this work. While this describes the types of problems that are tackled by researchers in EC, the variable elements within these categories can be broken down into the various types of change that can occur.

Table 3.1 Trojanowski and Michalewicz categorisation of problems split in two dimensions: objective function and constraints

No.	Objective Function	Constraints
1	static	null
2	static	static
3	static	var
4	var	null
5	var	static
6	var	var

Categories of Change

In comparison to categorisation of dynamic problems, categorisation of change has received more attention in the literature. De Jong derived a categorisation of different types of change in [52]. He describes four patterns of change:

i. Drifting landscapes: Here the landscape is undergoing small changes and it is the algorithm's job to track the optimum as it moves. An example under this category would be a manufacturing plant where the quality of materials may change over time, or machines may age and break down.
ii. The landscape undergoes morphological changes: This type of change includes landscapes where new regions of fitness may emerge where previously no optima resided. It is representative of competitive problems such as those seen in financial markets.
iii. Cyclic changes: This is where the change has an element of periodicity or the optima continuously reside within a certain region of the solution space. Problems under this category may include weather or political analysis.
iv. Abrupt and discontinuous changes: De Jong describes these changes as ones that may be cataclysmic and sudden, such as a power plant failure.

Branke too develops his own categorisation of types of change in [32, 33] that differ from De Jong's. He too describes four different types:

i. Frequency of change: How often does the solution landscape change?
ii. Severity of change: How far is it from the old optimum to the new?
iii. Predictability of change: Are the changes random or do they follow a pattern?
iv. Cycle length/accuracy: How accurate, and what is the period of a cycle in the solution landscape?

A final categorisation is presented by T&M [215] who propose three types of change:

i. Random changes: This describes changes that are not dependent upon the previous change or state.

ii. Non-random and non-predictable changes: These are changes that do depend on prior state and changes, but that occur as a result of relationships too complex to predict.
iii. Non-random and predictable changes: Here changes may be cyclical or follow some predictable pattern.

Of these three categorisations a certain amount of crossover and differences can be observed, with T&M presenting the most succinct classification. The T&M classification most closely constructs a taxonomy with the discriminatory factor being the degree of correlation in changes in the fitness landscape over time. Distinguishing types of change in this manner is useful as the EAs being developed will be expected to adapt their state or populations to these changes. Three different subgroups are identified by T&M. The first where the changes in the fitness landscape are Markovian in nature–entirely uncorrelated. The third where there is a predictable pattern to the change, where a correlation in changes in the landscape occurs over time and a deterministic prediction may be made. The second subgroup then is everything in between. This subgroup is large and may be refined into smaller subgroups to take into account endogenous and exogenous fitness functions. That is where algorithms take action as a result of their model or prediction and the consequences of the actions taken by the EA may not directly affect the changes occurring in the fitness landscape, or they do. The first and last subgroups only encompass exogenous fitness functions.

All three categorisations identify cyclical changes or changes that follow a pattern. De Jong and T&M could also be said to agree with regard to their identification of random changes; pattern 4 from De Jong with type 1 from T&M. De Jong describes pattern 4 as being abrupt and discontinuous indicating that it may be a change that would not have a dependency on the prior state and not leak information in the lead up to the change. Thus this satisfies the description given by T&M for their first type. This then leaves the non-random and non-predictable type of change from the T&M taxonomy and drifting landscape and morphological landscape changes from De Jong. Of these, the non-random and non-predictable type readily matches up with the morphological landscape given the example of financial markets provided by De Jong to describe this category. The final remaining type from De Jong can either be subsumed into the first or third types defined by T&M or fall somewhere in between the two. If the landscape is gradually shifting, with information on the shift's vector being leaked, then there are potentially predictive indicators that can be evolved placing this in T&M's third category. On the other hand, if information is not being yielded from the system then these gradual shifts are essentially the random changes identified in T&M's first category.

In examining the first two points from Branke, it is difficult to view frequency of change and severity of change as *types* of change. It is better rather to place them under the properties of a change that can be applied across all types of change, as they are measures of the rate and degree of the change. T&M also suggest that change can be examined under the criteria of being discrete or continuous but this basically describes the rate at which the change is occurring, matching up with Branke's frequency description. It should be noted that even if the rate of change is continuous, for the purposes of analysis, it will be required that the data be discretised on some level so as to allow evolution to occur on a stable data set.

In unifying these different categorisations, the opportunity is then presented to provide new and more widely recognisable names for the various types of changes that occur. T&M's first category "Random Changes" essentially describes a Markovian process where the changes that occur are independent of what has preceded. Such problems may include, for example, a sudden major disruption to infrastructure in a job scheduling system. T&M's second category also matches with De Jong's second and describes a Complex System where the changes that occur are a result of relationships too complex to accurately predict, for example, where multiple parties are competing against each other with various different goals such as those in the financial domain. As already stated this category includes fitness functions that may be exogenous or endogenous in nature. An example of such a change from the financial domain that includes an exogenous fitness function is fund replication where a risk portability profile is developed to replicate the returns of a fund with more liquid securities. An example with an endogenous fitness function is the act of investing itself and attempting to uncover inefficiencies in the market. The final classification from T&M, "Non-Random and Predictable Changes" suggests that there is information available in the build-up to a change that makes it possible to explain and predict the change. The change may occur as part of a cycle, a pattern, or some incremental process, because there are dependency relationships across the variables. As a result problems with this type of change are essentially Deterministic, where the current state is dependent upon that which has gone before and past information can be used to predict future states; analysis of weather may be considered under this category. Table 3.2 displays the results of this unification.

The properties of change described above and others can then be applied in varying degrees to each type of change listed in Table 3.2. This table will be used through out this chapter to identify what issues the various authors are seeking to address in their work.

Table 3.2 Unification of De Jong, Branke and Trojanowski and Michalewicz categorisations of types of change. The unified names are given in the first column followed by the mapping from each of the authors.

	De Jong	Branke	T & M
Markov	1 & 4	3	1
Complex System	2	-	2
Deterministic	1 & 3	3 & 4	3

3.2.3 EC in Static Versus Dynamic Environments

The question to consider here is as follows: what is the essential difference between evolving solutions for a static environment versus a dynamic environment? Traditionally GAs and other EC algorithms have worked on problems where the goal has been to deterministically solve a particular problem or reach a close-to-optimal state. With regards to dynamic environments this focus shifts–the emphasis is no longer on an individual becoming a specialist or expert at a certain task. Instead despite changes in an individual's environment, the focus is to track the optimum point as closely as possible, as stated and demonstrated by Huang and Rocha [95]. In essence, the aim of an individual in a dynamic environment is to survive. In order to do so it must develop a robustness or *"plasticity"*, as described by Rand and Riolo [178], despite the fact that the changes listed in the previous section obfuscate this goal.

Branke observes that the primary issue with Evolutionary Algorithms is their tendency to *"... eventually converge to an optimum and thereby lose their diversity necessary for efficiently exploring the search space and consequently their ability to adapt to a change in the environment when such a change occurs ..."*. Yet if the inspiration behind GAs is evolution, which works in an inherently dynamic environment [178, 121], it is difficult to see how the application of artificially simulated evolution to problems is itself misguided. The problem therefore must lie in our interpretation or implementation of evolution. De Jong notes that researchers often *"over fit (their) algorithms to various classes of static optimisation problems with the focus on getting the representation and operators to produce rapid convergence to near optimal points"*. The next section will explore the avenues researchers have explored in attempting to bring the results of EAs more in line with those observed in nature: adding extra features to prevent a telescoping of diversity and making the algorithms more amenable to changes in the problem landscape.

3.3 Existing Approaches for Evolution in Dynamic Environments

3.3.1 Overview

Standard GA and GP have led to 36 human-competitive results as defined in [112] in a number of areas [75] with complexity scaling up to a billion-variable problem [80]. This clearly demonstrates the ability of an artificial-evolutionary process to produce solutions through the stochastic recombination and mutation of solutions deemed to be fit in the problem environment. When this problem environment becomes dynamic however, as far as the authors are aware, there are no published results that are comparative to the patentable works cited for static environments.

As already stated populations in GAs exhibit a tendency to converge to an optimum with a resulting loss in diversity. This impedes the algorithm from efficiently exploring new areas in the solution space when the optimum shifts. This section looks at approaches researchers have adopted in overcoming this issue. Five approaches are examined:

i. **Memory:** Equip algorithms with a mechanism to recall previously effective solutions.
ii. **Diversity:** Prevent the noted problem of convergence within a population, resulting in the algorithm's failure to discover new solutions.
iii. **Multiple Populations:** Strike a balance between exploration and exploitation by assigning sub-populations to specific areas of the search space.
iv. **Problem Decomposition:** Break the problem down to its fundamental pieces.
v. **Evolvability:** Provide a representation that aids the population in producing offspring that are fitter than their parents.

Each approach will first be described, then analysed.

3.3.2 Memory

Of all the approaches investigated in attempting to make Evolutionary Algorithms' performance in dynamic environments comparable to that of natural evolution, the incorporation of memory into the algorithm has received the most interest [12, 29, 61, 100, 132, 131, 205, 214, 237, 238, 239, 240, 50, 49, 78, 82, 126, 144, 177, 189, 206, 95]. Within the application of this approach, Branke identified two distinct implementations [28]: Explicit Memory and Implicit Memory. Both of these are discussed next with respect to dynamic environments.

Explicit Memory: Background

Explicit memory involves the deliberate storing of genetic material in a memory cache. The philosophy behind this memory is that if a solution was useful in the past it may prove to be useful again at some point in the future. Various different strategies are employed in the selection, storage, and deletion of excess individuals to and from memory.

Louis and Xu [126] apply a memory feature to the shop scheduling problem, whereby at regular intervals of generations the best individual is stored. When the scheduling environment changes, for example, a faster machine is added or another machine breaks down, a new population is generated. However, 5–10% of this new population is seeded with individuals found in memory. The authors reported positive results with this level of seeding though saw a deterioration in performance due to either convergence with greater percentages or if the scheduling problem changed significantly (for example, if one job was deleted). Karaman et al. [100] also reinitialise their population, upon observing a change, with individuals from a stored memory.

With Ramsey and Grefenstette [177] the best performing individuals are again stored. In this case Ramsey and Grefenstette store the best-performing individuals in context at each change in the environment. They do this by measuring the state of the environment that the individual is being evolved and tested in. The state of the environment is measured through the use of a monitor module that captures the observeable parameters. When a change in the environment is recorded the population is reinitialised and 50% of the population is seeded with individuals from memory that have evolved in contexts similar to the current environment. Similarity is calculated by measuring the Euclidean distance between the observable parameters of the new environment and the previously recorded environments. Eggermont et al. [61] adopt a similar strategy and develop a case-based memory, which aims to keep track of interesting historical events and store individuals around these events. They test their solution on a periodically changing fitness function and reintroduce the stored individuals by randomly seeding the population after a change or, alternatively, by replacing the weakest individuals with memory individuals evolved in a similar context. Simoes and Costa [205] adapt the traditional GA paradigm to behave more like an immune system. The Immune System GA employs a memory population that attempts to remember prior pathogens and then selects the antibody cells that are most appropriate to battle against the pathogen in a process known as secondary response.

Mori et al. [132, 131] explore the use of a Thermodynamical Genetic Algorithm where at each generation the best individual is stored using a replacement strategy for the memory. The aim is to maintain a certain level of diversity, where diversity is the variance of bits in positions. This strategy attempts to maintain a wide representation of individuals across the solution space.

Branke [29] also follows the idea of maintaining a level of diversity in memory and adopts a dual-population paradigm. In this algorithm one population is re-initialised upon a change in the environment and the second population draws from a memory cache. The goal is to allow one population to explore potentially new areas of the solution landscape while the memory initialised population conducts search over previously discovered optima.

Bendsten and Kirk [12] create a dynamic model of memory. In their work an explicit memory population is initialised in parallel with the original population. At each iteration the individual from the memory population that is closest to the best performing individual from the solution population is selected and modified slightly to bring it closer to the solution individual on the solution landscape. This approach avoids the problems faced by others of adding to and selecting from the memory cache.

Trojanowski et al. [214] use a finite-size memory for each individual that stores successful ancestors in a FIFO queue. The better parent at each recombination is stored and the offspring inherits the memory of its parents. The memory is then employed when a previously unencountered obstacle is discovered.

Yang [237] employs a memory-based emigrants scheme, whereby the best solution in memory is used to seed new offspring and replace the weakest-performing members in the solution population. In further studies [238, 239, 240] Yang explores other forms of explicit memory. In conjunction with different forms of GAs, a population-based incremental-learning scheme stores the individuals working vector as well as the individual and an associative memory scheme.

Sebag et al. [198] introduce a negative memory with the concept of the *virtual loser*. The virtual loser is constructed by identifying the bits in the worst performers that are different from the best performers. Once this is done a flee-mutation operator is applied to new offspring that attempts to move the new individuals away from the virtual loser.

Explicit Memory: Analysis

The above approaches to the incorporation of memory into the evolutionary process are founded on the principle that it may be useful to remember previously successful solutions. Various different storage, selection and replacement schemes are explored and positive results are generally reported over a standard GA. However, problems examined also incorporate some element of periodicity, which is certain to favour any scheme that remembers previously successful solutions. The addition of memory does allow the algorithm to efficiently retrieve a previously successful solution. Through recombination with memorised individuals, or the seeding of new individuals with the genetic material of memorised solutions, it does provide a mechanism for shifting the population towards the neighbourhood of a previously discovered optimum. However this has been pointed out by Rand and Riolo [178] to have a negative effect in some cases, as it is based on the questionable premise that in a dynamic environment previous solutions will be useful in the future.

This addition of memory also comes at a cost in terms of computational effort due to the extra storage size of the memory individuals, which is generally the same size as that of the solution population. Derived from this, are the added evaluation costs of examining the memory population when a change occurs, measuring the environment, checking this metric against the stored contexts, and processing the steps involved in executing the chosen memory replacement strategy.

The intuitive use of a memory of previously discovered optimal solutions is appealing when considered for certain problems. Where a problem solution landscape is tightly bounded or undergoes a level of periodicity, the incorporation of an explicit memory, as shown by the above researchers, provides an efficient path to the rediscovery of optimal solutions or solution neighbourhoods.

The concept of an external explicit memory, however, does not readily exist in the natural evolutionary process. Though the study by Simoes and Costa [205] does present a parallel however, by placing the GA in the context of an immune system. The purpose of an immune system is to recognise non-self cells and mark them for destruction. The human immune system is made up of two parts; an *innate* and an *adaptive*. The innate part is encoded in our genome and is static from birth until death. The adaptive part of the immune system is modified each time it encounters a new pathogen, so that it is able to recognise this pathogen again in the future [24]. However, during this process no ontogenetic process occurs and the patterns recognised by a parent's adaptive immune system are not be passed onto his/her offspring.

Implicit Memory: Background

Implicit memory differs from explicit memory in that it does not utilise an external memory cache of previous solutions or states, negating the need for extra selection, replacement, and storage strategies. Implicit memory instead is incorporated into the individuals of the population through degenerate genetic material. A number of authors have adopted different approaches in the creation and utilisation of implicit memory though most researchers focus on the development of multiploidy representations. Multiploidy is where more than one allele is used to represent each individual. Whereas in standard evolutionary computation representation one allele is used to represent the solution, this is known as haploid representation. Multiploidy includes a multitude of alleles to express a phenotype and diploid is where just two alleles are used. The use of a multiploidy representation requires some mechanism to enable the dominance of one allele over the other. The literature expresses almost as many dominance mechanisms as researchers investigating the mechanism.

Goldberg and Smith [78], and Smith [206] produced the earliest publications on a diploid representation for improved evolutionary results in non-static problems. Their results demonstrated that a diploid structure with an evolving dominance map provides superior results over a haploid representation. They show that the diploid structure is able to maintain alternating alleles in abeyance, thus affording a quicker evolutionary process to an alternative optima. Hadad and Eick [82], and Ng and Wong [144] also present diploidy representations with varying dominance mechanisms.

Dasgupta and McGregor [49, 49] implement a multiploidy approach where genes at higher levels are able to activate or deactivate the genes at the immediate lower level enabling a hierarchical structure to be evolved.

Ryan [189] develops a multiploidy representation where the dominance is performed through the summing of genes for a particular trait. If the result exceeds a certain threshold the phenotypic trait is 1, or 0 otherwise. This methodology was extended in Ryan and Collins [191], by introducing a lower threshold. Where the summation was below this the phenotypic trait was 0, in between the two bands the trait was generated randomly, and above the higher threshold it was 1, as in the previous study. This methodology produced superior results to the prior study.

Osmera et al. [170] found that for highly multi-modal problems the diploid structures were able to outperform their haploid counterparts. Dominance in this case was performed through an XOR operation between chromosomes. Callabretta et al. [37], similar to Osmera, also employ an XOR mechanism. In this case the first bits of the alleles are XOR-ed. If the result is 0, then the allele with the second bit equal to 0 is expressed, with the same algorithm being used if the result of the XOR is 1. If the second bit in both is the same, then the alleles are co-dominant and an average between them is used.

Collingwood et al. [45] adopt a mask to determine dominance. This specifices which of the chromosomes is dominant independent of which allele is present in that chromosome. Each individual in the population possesses its own mask that is also evolved through the application of genetic operators during reproduction.

Kim et al. [243] use a *winner take all* approach whereby each chromosome is evaluated with the phenotype being expressed, derived from the fitter chromosome. In this case reproduction occurs among the dominant chromosomes of parents to generate one offspring, and the recessive chromosomes to generate another.

In, [223], Uyar and Harmanci conduct an analysis of performance of different genotype-to-phenotype mapping schemes. However, having completed this survey unsatisfied they adopt their own dominance mechanism [224] that aims to maintain a balance between exploration and exploitation. It does this through adaptation of the dominance mechanism as a result of feedback from current phenotypic output. Lewis [119] also conducts a survey of multiploidy techniques and concludes that a simple multiploidy scheme does not provide a strong advantage over a haploid GA. When equipped with a dominance mechanism, multiploidy algorithms do perform better though similar to a haploid GA, which throttles mutation when a degradation in fitness is observed. For diploid representations, such as that presented by Ng [144], Lewis comments that their utility is largely limited to functions that oscillate between two optima.

Though the exploration of multiploidy and dominance mechanisms has received the majority of attention from researchers for the incorporation of degenerate genetic material other methods also have been examined.

Ohkura and Ueda [165, 166], present a string representation that contains inactive regions. These inactive regions allow for the possibility of neutral mutations to occur. Over the course of the evolutionary process active and inactive regions will change. They will present a higher level of diversity across the population along with the incorporation of memory in inactive regions that can be reactivated by evolution when a change occurs.

Huang and Rocha [93, 94, 95, 181, 182] explore the use of RNA editors in their studies. RNA editing occurs during the mapping process from genotype-to-phenotype and applies stochastic variations to this process. This results in genotypically equal candidates producing different phenotypes, meaning a many-to-one mapping from genotype-to-phenotype and vice versa. RNA editing in nature occurs in the development process of an organism and so is not ontogenetic, meaning the changes that occur in the mapping process are not inherited by offspring. Huang and Rocha report greater plasticity and robustness of solutions as a result of the RNA editing over traditional GAs. The RNA editors can be classified as implicit memory because in [95] they develop an agent-based model where each individual evolves and possesses its own RNA editor.

In [148], O'Neill highlights the presence of degenerate genetic material in Grammatical Evolution. In this case a genotype-to-phenotype mapping occurs that involves a many-to-one relationship and, like the studies by Ohkura and Ueda, the ability to conduct neutral mutations is also present. Though GE has largely only been tested on static problems, with the exception of [156], considering the prior studies examined here it would suggest that these features give GE a plasticity of solution. The genotype-to-phenotype mapping in GE is discussed in greater detail in Chapter 2.

Implicit Memory: Analysis

The concept of diploid or multiploidy in EC is inspired by similar representations in the natural world. Each human cell contains 23 pairs of chromosomes, one from each parent. This cell structure is considered to be diploid. Different strands of chromosomes then, have the attribute of being either dominant or recessive and it is this relationship, for example, that dictates whether we have brown or blue eyes.

In surveying the work to-date on exploring the potential implementations of Implicit Memory, it is clear that a multiploidy representation has motivated the majority of researchers' efforts with mixed results. Comparative studies on multiploidy representations with different dominance schemes have demonstrated that their performance is equivalent to that of haploid GAs with higher or adaptable mutation rates. The central criticism of Lewis' study is that the multiploidy representations he examined were essentially only useful for problems that alternated between a couple of solutions [119], while Branke also makes a similar observation [27].

While this problem could also be considered an advantage in applying the *"horses for courses"* addage – in that there are certainly problems in the real-world to which a diploid structure may be applied quite successfully – there is still the problem of selecting a suitable dominance function. Of the work surveyed, no clear solution was presented that had received any degree of traction. This is evidenced by the diversity of dominance functions. It should also be noted that the representation of multiple versions of the genotype for each individual also comes at a computational cost.

Ohkura and Ueda, and Huang and Rocha provide alternative implementations of Implicit Memory, where redundant genetic material is not stored in a diploid or multiploid structure thereby negating the need for a dominance function. With Ohkura and Ueda, the degenerate genetic material is represented by inactive regions on the genotype presenting a resilience to destructive mutations, while also providing memory when those regions are reactivated. Huang and Rocha develop an Implicit Memory mechanism with the addition of stochastic RNA editors. These editors allow a single individual to present multiple phenotypes each time the RNA editor is run, favouring one phenotype over the other by adapting the probabilities of the rules the RNA editors use.

While the representations expressed by researchers in the previous section do have a strong foundation in the natural world, the development of the various dominance functions does not mirror the natural parallel to the same degree. This may be partly down to the complexity of dominance in the natural world where there may in some cases be partial dominance or also co-dominance. Such variations present challenges when attempting to implement the relationship in algorithmic form.

Of course, what Explicit Memory and the Implicit Memory outlined here do not cover is the fact that there is also an Implicit Memory inherent in the population as a whole. The memory resident in the population itself becomes more important in dynamic environments, especially where changes are of a more gradual degree. Under such circumstances previously optimal solutions may slip in the rankings and stand less of a chance of passing on their genetic material once the optima have moved. However, they do provide the population with a means to quickly reproduce older useful solutions if the environment reaches a state similar to one previously visited.

Reflecting on the categories of change identified earlier in Table 3.2, it can be seen that through the incorporation of memory into the GA, the authors were seeking to address two types of change, Complex and Deterministic. In the case of Complex changes memory of past events can prove to be useful as they may give insights into future changes where similar conditions exist. Where the changes are Deterministic in nature, cycles or patterns may exist and previously discovered solutions can be readily recalled. In situations where the type of change is Markov, memory does not play as important a role as these changes are random and remembering previous states or good solutions will not aid in an algorithm's search.

Table 3.3 The addition of memory is aimed at helping systems remember past useful solutions for problems where there may be cycles or patterns

Problem Type	Memory
Markov	-
Complex System	X
Deterministic	X

3.3.3 Diversity

GAs can be considered to be parallel problem solvers, where a population of solutions spread out across the solution space attempt to find and converge to a global optima. While in the initial dispersed state, a wide coverage of the solution space is achieved with many local optima being explored. However, as the population converges, this search narrows in focus. This problem has been alluded to by a number of authors [32, 205, 74, 81] especially with respect to dynamic problems. Diversity itself can be measured in a number of ways [36]: through explicitly counting the number of distinct genotypes, phenotypes, and fitnesses; using entropy [185]; or implicitly, by monitoring the time-averaged-fitness of the population and associating a drop in this average with a narrowing of diversity.

Diversity: Background

As mentioned above, a common criticism of GAs is that the population converges to an optimum, leading to a drop in diversity and making the discovery of new optima difficult when the environment changes. Increasing the level of diversity after a change has occurred, or been noticed, is one way of addressing this issue. However, this fails to take into account the clustering of a diverse population and so dispersion of the population must also be considered.

The most popular solution is *hypermutation*, proposed by Cobb [44], where the rate of application of the mutation operator is increased at certain times. Cobb uses hypermutation to maintain an acceptable level of time-averaged best performance. When this average performance drops below a threshold, hypermutation is applied with the aim of maintaining a balance between exploration and exploitation. The implication of this increased rate of mutation is that a higher level of diversity is achieved as opposed to what could be achieved through crossover alone. The issue being addressed by hypermutation is that in a converged population, crossover between parents fails to increase diversity leading to the need for greater mutation.

Vavak et al. [225] also examine the time-averaged best performance as the trigger for the application of their *variable local search* (VLS). When this occurs, crossover and mutation are temporarily suspended. VLS is conducted by applying a shift bit register to the individuals selected for reproduction. The bit register is made up of bits that are set randomly and the value represented by this register is either added to or subtracted from the bit string of the individual. This enables a local search around the location in the search space represented by the solution. The register is applied to each member of the population and then the GA resumes with its normal operators. If performance is not improved after a number of generations and/or a period of time, the VLS is again applied with an extended area of search, in other words, the register is longer allowing a wider boundary. In [226], Vavak et al.

extend their study of VLS with the introduction of an incremental learning technique for controlling the boundary of search automatically.

As an alternative to the monitoring of time-averaged fitness, the explicit maintenance of diversity over the course of evolution also addresses the issue. Mauldin [128] presents an early example of the maintenance of diversity for static environments through setting an adaptive threshold of Hamming Distance between bit strings for new individuals to be introduced to the population, with this threshold decreasing as generations progress. Mauldin observed that this methodology produced more robust solutions when applied to the out-of-sample data.

Grefenstette [81] maintains diversity by introducing the concept of replacing a certain precentage of the population at each generation with randomly generated individuals, a process that has become known as *random immigrants*. This process maintains a level of diversity throughout a run by potentially introducing new genetic material to the population at each generation.

Cedeno and Vemuri [39] employ a *Multi-Niche Crowding GA* that promotes reproduction among solutions sharing similar traits, which collectively could be described as occupying a niche. The replacement strategy that is then adopted replaces the worst-performing individuals in this niche. This has the effect of essentially preventing a speedy convergence in the search space by encouraging similar solutions to reproduce in their niche.

Ghosh et al. [77] keep track of the ages of individuals in the population. In doing so they are enabling a selection bias based on the age of the individual in conjunction with its score derived from the objective function. The age-scoring function adopted in this study intuitively favours the middle aged over the young or old. Such a scheme provides less of a chance for selection of poor-performing young and old solutions.

Mori et al. [132, 133] aim to maintain a temperature or a level of free energy in the evolution of a population through the use of a Thermodynamical Genetic Algorithm (TDGA). This algorithm maintains this level with the expression

$$F = E - TH \tag{3.1}$$

where F is the energy level of the population, E is the average fitness of the population, H is the diversity of the population and T is the weighting placed on diversity. Individuals are introduced into the population after reproduction in such a way as to maintain a high level of F.

Simoes and Costa [204] develop a new genetic operator known as a *transformation* whereby genetic segments from previous solutions, which have left the population, are stored in a pool. These segments are then inserted into individuals at reproduction replacing the crossover genetic operator and avoiding the problem of crossover in a converged population.

In [135] Morrison employs *sentinels*, which are individuals uniformally dispersed [134] in the search space. These sentinels remain statically located and play normal roles in selection and reproduction, while also providing the algorithm with the ability to maintain a level of diversity that is dispersed throughout the search space. When a radical shift in the environment causes the optimum to move to a different area, a sentinel will already be located in the general neighbourhood to provide genetic material that will help in moving the general population towards this new position.

Diversity: Analysis

Reflecting on nature where almost every animal on the planet possesses its own unique genetic makeup, diversity is seen to play a central role in the natural environment. It is this diversity that enables populations to produce solutions to new diseases or changes in their environment, where each problem is tackled in a massively parallel manner through a multitude of unique genetic makeups. In contrast, studies have shown [231, 127] that inbreeding and a reduction in diversity can have a negative effect on the fitness of organisms in nature, lending credence to approaches that place an importance on diversity in evolutionary algorithms.

Branke [27] identifies two ways in which researchers use diversity as an aid to EAs for dynamic environments: to increase diversity after a change has occurred and to maintain a level of diversity throughout the run. The maintenance of diversity can be further broken down into the active maintenance of diversity and the pre-emptive maintenance of diversity. Both Cobb and Vavak's solutions fall under Branke's first classification. In their studies, a time-averaged drop in fitness is used as an assumption that the population has converged and is struggling to explore new solutions. This approach, however, may suffer in real-world applications. The reason is that if it has become noticeable that the time-averaged fitness has dropped below a certain threshold, the system is already failing or underperforming in its task. Also implicitly associating fitness with diversity, instead of monitoring or maintaining it directly, may mask other issues. The manner in which these two solutions are applied, whereby the degree or rate of application is increased with a drop in fitness, suggests that these solutions are essentially temporary fixes or bandages on the more fundamental problem of maintaining diversity and exploring new areas of the search space.

Maintaining a level of diversity throughout the life cycle of a system certainly addresses these problems. Simoes and Costa, and Grefenstette continuously introduce new genetic material to the population by recycling extinct genes and introducing newly generated random material. Such mechanisms are necessary to prevent the problem described earlier, whereby crossover in a converged population does not yield new genetic material. Ghosh et al., and Mori pre-emptively solve the problem through selection and replacement strategies that attempt to prevent any individual from taking too strong a hold of the population and dominating the genetic pool. Morrison too pre-emptively maintains diversity in his contribution, but more than this Morrison also ensures the population maintains a level of dispersion throughout the search space. This addition prevents a population, which may be genetically diverse, from clustering around one area of the search space and therefore provides the algorithm with the ability to maintain a dispersed coverage of a bounded search space. Table 3.4 summarises the classifications.

Table 3.4 Authors and the three approaches to diversity: increasing diversity, actively maintaining diversity, and pre-emptively maintaining diversity

Increase	**Maintain Active**	**Maintain Pre-emptive**
Cobb	Simoes & Costa	Ghosh
Vavak	Grefenstette	Mori
-	-	Morrison

This considered, an issue that largely escapes attention in the EC literature is whether in fact the drop in diversity is a result of too much evolution being conducted. Branke [33] attempted to measure change severity that could potentially have led to an evolution-throttling mechanism, though the developed metric proved to be inaccurate. Dempster and Jones [57] monitor for convergence in their population and stop evolution when the fitness of a number of the most fit individuals fails to improve by no more than 1%. This issue is potentially a legacy problem from static environments, where evolution is used as an optimiser and in a sense to overfit to a particular problem.

Ensuring a population is adequately diverse aids the algorithm in discovering new solutions in areas of the search space that before had provided poorly fit individuals. In continuing the development of Table 3.3, diversity is tailored to aid in the navigation of two types of changes, Markov and Complex Systems. For Markov type changes maintaining a dispersed and diverse population can ensure that some member will be in a relatively fit neighbourhood when a change occurs. With complex types of change the model of change is unknown and is also subject to change. Maintaining a diverse population in this case enables the system to discover relatively fit neighbourhoods as the optima move and also to evolve innovative solutions from areas of the solution space that had not previously been explored. Deterministic problems in this case are not specifically catered to. Yes, as in all problems, a level of diversity is necessary to adequately explore the solution space. However, in this case a converged solution is sought in uncovering underlying pattern or structure and the maintenance of a diverse population carries less weight.

Table 3.5 Diversity enables the algorithm to discover new solutions in Markov and Complex System problems

Type	Memory	Diversity
Markov	-	X
Complex System	X	X
Deterministic	X	-

3.3.4 Multiple Populations

Again addressing the issue of population convergence, researchers have adopted a separate strategy different to that of maintaining diversity in a population through the creation of multiple populations by partitioning the main population. The aim here being that separate multiple populations can exploit discrete areas of the solution landscape. This is based on Wright's (1932) Shifting Balance Theory (SBT). In this theory, Wright recognised that within a population premature genetic convergence can occur. However, if the population is split up into smaller subpopulations, genetic drift allows these discrete subpopulations to exploit new areas on the fitness landscape. A GA first developed along the lines of this theory is known as the Forking GA and was first developed by Tsutsui et al. [218] for static environments.

Multiple Populations: Background

Oppacher and Wineberg [167] developed the first SBT inspired GA for dynamic environments. Key to the implementation of such a system is the incorporation of a mechanism for meaningfully dividing the subpopulations. Oppacher and Wineberg achieve this through use of Hamming Distance. If a particular colony population overlaps the core population, extra evolutionary pressures are activated to push that colony away from the core population. Migration is then conducted from the colonies to the core population in a stochastic manner, based on a colony's fitness. In this case, the core population is responsible for the exploitation of best solutions, while the colonies explore the landscape for new promising optima.

Branke et al. [30] approach the problem from a slightly different angle through the Self-Organising Scouts model. Here, the larger base population searches the landscape for new undiscovered optima and assigns a scout subpopulation to areas which show promise. At each generation, an analysis of the population is conducted in order to determine whether there exists a group of individuals that are clustering around a phenotypic optimum. If this is the case, this group is separated from the base population and undergoes evolution only within the group to fully exploit the area.

Multinational GAs developed by Ursem [222] identify groups of individuals attempting to form nations. A nation is formed when a fitness valley exists between the most fit individual in an old nation and the most fit of a potentially new nation, calculated by evaluating random points between the two individuals. A variable mutation rate is also adopted within nations where individuals on the periphery undergo a higher rate of mutation compared to the most fit individuals within the nation. While this method did provide evidence of being able to track multiple optima, it suffered from a lack of diversity on a global level.

Another method of splitting up the population is explored by Ronnewinkel and Martinez [184] based on a fitness envelope defined by Petrowski and Genet [171] where fitness envelopes are formed around centroids. This method has benefits over the previous paradigms as it does not require extra fitness evaluations to determine valleys and also does not require parameters for maximal distance around a centroid to encompass its colony.

Multiple Populations: Analysis

Through splitting up the population into subpopulations charged with exploring different areas of the fitness landscape, a GA can exploit multiple optima simultaneously. The number of potential optima to be explored, however, is limited by the minimum subpopulation size and the global population size. As demonstrated by Ursem, diversity can still become an issue with the population potentially converging to a range represented by converged subpopulations.

The function of multiple populations is to strike a balance between exploration and exploitation. In this regard its intention is generally applicable to search and as such is not catering to specific types of change in dynamic environments. Indeed, its benefits also extend to static environments.

Table 3.6 Using multiple populations aids search in general as it tries to strike a balance between exploration and exploitation

Type	Memory	Diversity	Multiple Populations
Markov	-	X	X
Complex System	X	X	X
Deterministic	X	-	X

3.3.5 Problem Decomposition

The act of problem decomposition is the breaking up of a problem into its more fundamental constituents. This relates to the building block hypothesis [89, 79], which states that GAs work well when short, low-order, highly-fit schemata or building blocks, recombine to form even more fit higher-order schemas. While analysis of this theory has largely been conducted for static environments [227, 69, 208], its importance is also carried through to dynamic environments where the formation of robust generalised solutions goes hand in hand with problem decomposition.

Problem Decomposition: Background

Abass et al. [1] approach the problem of evolution in dynamic environments by assuming that the most efficient way of tracking an optimum in a changed environment is to extract structural knowledge about the problem from previously successful solutions. Using the information theoretic measure, the minimum description length (MDL) [180], a probabilistic decomposition model is established for the structure of the problem and manipulated as the environment changes. This model is then used to identify building blocks and to conduct crossover along the partitions of these building blocks. In their work Abass et al. maintain the decomposition model across changes in the environment, but restart their population of solutions at each change.

In Ronnewinkel et al. [183] a theoretical examination of a GA is undertaken, with mutation as the only genetic operator. Where regular changes occur in the environment, defined as changes that have a fixed duration and obey some deterministic rule that is the same for all change cycles, Ronnewinkel et al. observe the correspondence of their model to Eigen's *quasi-species* [62]. A quasi-species is a generalised self replicating entity that can be represented by a small number of building blocks. New genetic material is arrived at through mutation, which occurs during the copying process. The identification of the quasi-species model in a GA provides evidence of problem decomposition by identifying atomic problem building blocks that allow the algorithm to more efficiently adapt to changes within the problem's environment.

Problem Decomposition: Analysis

While the majority of research conducted on the building-block hypothesis and problem decomposition has focused on static environments, its use and understanding is of equal or greater importance for the construction of robust solutions for dynamic environments. Eigen's evidence of problem decomposition in the form of the quasi-species in a natural setting reinforces the importance.

The identification of a quasi-species and building blocks for problem decomposition also ties into one of the approaches discussed earlier: memory, or implicit memory more exactly. As a population evolves and begins to learn the structure of a problem, this information gets incorporated into the genetic structure of the individuals. Therefore, it operates as a type of memory and provides the population with a level of competency in its problem domain. This implicit memory, or population competency, then intuitively leads to the generation of solutions that will be more robust to changes.

For problem decomposition, the intention is to evolve a robust, competent population of solutions. This necessitates that the problem domain offer some information that may be analysed to provide an insight into how it is changing. In Table 3.7 problem decomposition is seen to aid with two problems. In Complex and Deterministic changes information may be extracted from the domain, providing insights to the nature of the changes or the structure of the solution. On the other hand, a system's response to a Markov type of change is essentially reactionary, as prior information or state does not aid in predicting the change.

Table 3.7 Problem decomposition requires the presence of information and dependencies in the domain

Type	Memory	Diversity	Multiple Populations	Problem Decomposition
Markov	-	X	X	-
Complex System	X	X	X	X
Deterministic	X	-	X	X

3.3.6 Evolvability

Where problem domains are dynamic, the ability of a population to evolve to a new area in the solution space becomes key. Navigating from a local optimum to a global optimum or following a moving optimum is more efficient when the representation or paradigm used makes this process easier. Providing a high level of evolvability may allow the migration to a new, better optimum without the population or individuals first experiencing a drop in fitness as they traverse the fitness landscape. In natural evolution, evolvability is described as the capacity for an adaptive response to a dynamic environment [83].

Evolvability: Background

Altenberg [3] approaches the issue of evolvability from a theoretical standpoint in a static environment and examines the *constructional fitness* of blocks of code. The constructional fitness is the probability that a block of code will increase an individuals fitness if it is added. A genetic operator then, based on this premise, will become focused on the dimensions of variability with a greater potential for increasing fitness. Altenberg recommends Upward-mobility selection (a type of steady state selection) and Soft Brood selection (a tournament selection is applied to the offspring before they are added to the population) as means of improving proliferation of constructionally fit blocks through the population. For recombination he advises the use of a type of genetic engineering where blocks of code are selected from a library which has a rated catalog of blocks based on their historical performance in increasing an individual's fitness after insertion.

Risinger et al. [179] note that in dynamic environments more evolvable representations are better able to survive in the long-term. With this in mind, they set out to measure the evolvability of three different representations on a dynamic fitness function: a direct encoding, a modified direct encoding with parameters affecting how mutations should occur, and Genetic Regulatory Network (GRN) that uses a genotype-to-phenotype representation. The GRN presented highest level of evolvability due to its ability to conduct neutral mutations which enabled it to maintain a more diverse population and more easily switch to a new solution.

Shipman et al. [201] seek to harness the benefits of evolvability through the use of a many-to-one genotype-to-phenotype mapping, which is applied to the real-world dynamic problem of planning for the growth of a telecommunications network. They observe that such a representation allows for the potential of neutral drift that can migrate a population out of a local optimum. The GRN representation they adopt allows this, as the evolutionary search can continue in the genotypic space through neutral mutations while the underlying fit phenotype is maintained.

In Ebner et al. [60] the work of Shipman et al. is extended through drawing a link between phenotypic performance of an indirect encoding and the existence of neutral networks in the genotypic space. These neutral networks are identified by Engelhart and Newman [140] in the natural world where neutral mutations allow the maintenance of genotypic diversity with genes spread out along a "neutral" network while still mapping to a valid fit phenotype. This provides the ability to easily evolve to a new, fitter phenotype when circumstances change as the search space widely covered at the genotypic level. Ebner et al. examine the reachability of phenotypes, the innovation rate, connectivity between phenotypes and the extent of neutral networks where a direct encoding representation is also used as comparison. Across all metrics the indirect encoding provided far superior results due to it possessing the ability to easily reach other phenotypes with small shifts in the genotypic space.

Evolvability: Analysis

Providing an evolutionary algorithm with a strong level of evolvability gives it the potential to incrementally improve its population fitness without first experiencing a decrease as it escapes a local optimum. Where the environment is dynamic a high level of evolvability allows the population to more easily track a moving optimum. Indeed, where the environment is dynamic, a high level of evolvability is rewarded in the evolutionary process, as it has a greater chance of survival.

Altenberg focused on a direct encoding in a static environment and as such the selection and recombination recommendations made suggest a high rate of convergence. Under such schemes it may be difficult to shake out of a neighbourhood where a target moves. However, in the representations examined above, those with an indirect mapping are implicitly coupled with another desirable attribute: a potentially higher level of diversity in the genotypic space. Because of the many-to-one mapping, individuals may spread out along a neutral network in the genotypic space providing the high levels of connectivity described by Ebner in the phenotypic space. Combined with this, such a representation also facilitates neutral mutations.

In adopting a many-to-one genotype-to-phenotype mapping a greater level of evolvability can be achieved which parallels features observed in nature. For dynamic environments the ability to easily reach another solution is imperative. Whether the type of change is Complex, Markov or Deterministic, providing good evolvability essentially lubricates the evolutionary process.

Table 3.8 Evolvability is key across all dynamic problems where the ability to adapt and improve fitness in a changing environment is necessary for survival

Type Type	Memory	Diversity	Multiple Populations	Problem Decomposition	Evolvability
Markov	-	X	X	-	X
Complex System	X	X	X	X	X
Deterministic	X	-	X	X	X

3.4 Evaluation of Performance

3.4.1 Evaluation of Performance: Problem Description

In static-environment problems, comparing the performance of different algorithms is not typically difficult. Comparison of final fitnesses and the rate at which they are achieved are readily available and serve both quantitative and qualitative purposes. Where the environment is time variant, the final fitness only represents one snapshot into the life cycle of the dynamic process; it does not capture the performance of the system over time.

For static environments Feng et al. [66] formalise a number of metrics examining:

- Optimality
- Accuracy
- Sensitivity
- Convergence and
- Optimiser overhead.

Of these, Optimality and Accuracy may be used as measurements of a dynamic algorithm's quality through taking snapshots at regular intervals. However, these measurements, as defined by Feng et al., assume that the global optimum is known and that the solution space is bounded. These two constraints may render the metrics unusable when applied to real-world dynamic problems.

3.4.2 Evaluation of Performance: Metrics for Dynamic Environments

De Jong [51] identifies two types of performance set in static environments: on-line performance and off-line performance. On-line performance is the performance or average fitness of the best solution during training on the in-sample data. Off-line performance is the fitness of the best solution on the out-of-sample data. Even though intended for static in-sample/out-of-sample tests, these types of performance bear relevance in dynamic environments where, during on-line training, the goal is to track the moving optima as closely as possible. Then when the system is live on new data, tracking the off-line performance becomes necessary.

Mori [132] derives a metric that produces the average ratio of best-evolved fitness to the global optima over time. Its equation is:

$$M = \frac{\sum_{t=1}^{T_{max}} \frac{f_{best}(t)}{f_{opt}(t)}}{T_{max}} \tag{3.2}$$

where T_{max} is the length of the search process or the number of time steps in the search, $f_{best}(t)$ is the best solution at time step t, and $f_{opt}(t)$ is the global optimum at time step t.

Trojanowski and Michalewicz [215] develop two metrics specifically for dynamic environments that measure accuracy and adaptability. The accuracy metric measures the difference between the current best solution before a change in the environment and the optimum value averaged over the entire run. It is given by the expression:

$$Accuracy = \frac{1}{K}\sum_{i=1}^{K}(err_i, \tau - 1) \tag{3.3}$$

While the adaptability metric measures the difference between the best individual at each generation and the optimum value averaged over the entire run, it is expressed as:

$$Adaptability = \frac{1}{K}\sum_{i=1}^{K}[\frac{1}{\tau}\sum_{j=0}^{\tau-1}(err_{i,j})] \tag{3.4}$$

where $err_{i,j}$ is the difference between the value of the current best individual of the jth generation after the last change and the optimum value after the

i^{th} change. τ is the number of generations between two changes and K is the number of changes of the fitness landscape during the run.

Besides these metrics, other measurements of performance have been developed by charting differences between best performing individuals over time relative to the global optimum at the same time.

3.4.3 Evaluation of Performance: Analysis

A common thread through these metrics is the incorporation of the global optimum into the calculations. Unfortunately, however, in real-world dynamic problems it is often the case that a non-deterministic problem is being analysed. Under such circumstances, the comparability of the metrics across different problems is not possible. Indeed, the utility of comparing accuracy and adaptability across different problems is indeed questionable in the first place. For example, results achieved in comparing a symbolic regression system on an unbounded random walk, and a simple linear function will produce vastly different performances.

In examining the performance of algorithms on real-world dynamic problems, often the domain in question will possesses a wealth of relevant metrics. Indeed, such metrics produce a more relevant analysis of performance as the results can be compared against acceptable norms in the domain or in the performance of human competitors. In the domain analysed in this book, which is finance, a wide variety of metrics are used to analyse both the performance and behaviour of the subject system.

3.5 Benchmark Problems

Benchmark problems are used to provide a common baseline in understanding the behaviour of new algorithms or extensions to existing algorithms. The problems are generally straightforward to reproduce and allow researchers to gain a relative perspective of their work compared to others. According to Branke [29], benchmarks should be:

- Easy to describe
- Easy to analyse
- Tunable in their parameters and
- Strike a balance between complexity and simplicity.

With the aim of satisfying these criteria, Branke introduced the Moving Peaks benchmark. This benchmark is similar to Morrison and De Jong's [134]. The Moving Peaks benchmark is a multivariate problem where multiple peaks of varying height and width shift and change their dimensions after a number of generations. The goal here is to locate the highest peak.

Yang [235, 236, 237] produces a dynamic benchmark suite that varies in difficulty using a decomposable trap function. This problem is also multi-variate, as the function decomposes to constituent building blocks and the difficulty of the problem can be changed on line.

Huang and Rocha [95] examine their system using an oscillating Royal Road benchmark along with trials on an Optimal Control Testbed, which crosses into the domain of engineering sciences.

Rand and Riolo [178] conduct benchmarking tests using Shaky Ladder Hyperplane-Defined functions that, like Yang's benchmark, are designed with building blocks in mind and are based on schemata.

de Franca et al. [70] develop a suite of dynamic problems based on the work of Angeline [6], where optima are displaced every n generations.

Many other researchers examine their algorithms on time-varying knapsack, oscillating, and scheduling problems.

3.5.1 Benchmark Problems: Analysis

Despite the best efforts and intentions of a number of authors, the number of *benchmark* problems in use, like dominance functions in Section 3.3.2, remains as varied as the number of researchers involved. This then has the negative consequence of placing the reader in the situation where benchmarks are being compared ahead of the actual algorithms.

Hidden by this problem is the question of what the researchers are actually aiming to achieve. If the long-term view is the application of algorithms to real-world problems, there exists a very real danger that algorithms are being optimised to solve benchmarks and not the relevant problem at hand. Added to this, a wide range of real-world problems from a multitude of domains exist and it is naive to assume that a particular benchmark can capture the various intricacies involved in each.

Benchmark problems certainly have their use as logical unit tests for extensions to algorithms in order to gain fundamental understanding of their behaviour. With this in mind, different benchmarks should simulate the various categories of change outlined in Section 3.2.

3.6 Chapter Review

The core aspects of EC in dynamic environments have been examined. The chapter began with an examination of dynamic environments and how they differ from static problems. This was followed by an attempt to classify and unify independent categorisations of the types of changes that occur while also identifying a number of properties of change.

Following from this an in-depth analysis was made of the various approaches adopted by researchers in advancing Evolutionary Algorithms to produce better performance in dynamic environments. The first approach analysed was the addition of explicit memory to algorithms. This generally involves the inclusion of a cache or extra population to serve as a memory for previously effective solutions. The philosophy being that these solutions may prove to be useful again if the environment returns to a similar state, or their genetic material can provide a source to help direct the search neighbourhood. While results presented by authors do demonstrate improvements in benchmarks, the approach comes at a computational cost and has the undertones of being a brute-force algorithm. Indeed it begs the question, why not store all good solutions? It also rests on the assumption that previously learned optima may be useful, or returned to, in the future.

Implicit memory differs from explicit in that information is stored in degenerate or redundant genetic material. The favoured implementation in the literature is through the incorporation of diploid or multiploid genetic structures for the individuals. This enables the individual to store previously good information and swap quickly when a change occurs. This paradigm was seen to be limited in its success. It is applicable predominantly to problems that oscillate between a small number of states while still incurring a computational cost, and does not demonstrate any clear cohesive approach to the subject of dominance functions. It is noteworthy that both the explicit and implicit multiploid forms of memory come at a computational cost when one of the aims of EC in dynamic environments is to reduce computation cost by adapting a population, rather than restarting it from scratch, with each change. However, still under the umbrella of implicit memory, flexible, computationally-efficient methods are demonstrated through the use of a many-to-one genotype-to-phenotype mapping process, along with the use of degenerate genetic material in a linear-string representation. These methods were found to be more adaptable to multiple states over a standard EA, while the efficient representation meant that memory was maintained at little computational cost. Such a representation with a many-to-one mapping also addresses Eggermont's criticism [61] that the population, which is supposed to contain the history of evolution, does not keep enough information to allow the algorithm to react adequately to changes.

In examining the role of diversity, two paradigms were also observed. Examined first was the increasing of diversity triggered by a drop in performance. This approach while serving the purpose of increasing diversity, suffers in application, where the lag inherent in the triggering mechanism can mean a period of sub-par performance before recovery. The second approach, simply maintaining the level of diversity, is achieved in the literature through the use of a number of different strategies for both actively maintaining the level of diversity and pre-emptively maintaining it. None of these strategies, however, look on the issue of convergence within the population as a representation problem or one of overfitting. By adopting a representation with a many-to-one genotype-to-phenotype mapping, the search space is separated from the solution space, allowing a much greater level of freedom in the genotype space.

By partitioning their populations, researchers use multiple populations to probe separate areas of the solution space simultaneously. In doing so, a balance between exploitation and exploration is sought, though limitations are encountered in large search spaces where a population can only be divided up a certain number of times. Added to this, a loss of diversity is observed without maintaining migration between the populations.

Problem decomposition and the building block-hypothesis were examined next. To a large extent this theory addresses the good intentions that inspired the incorporation of memory into GAs for dynamic environments. By seeking to identify the structural or theoretical nature of the problem, an algorithm is able to establish an understanding or identify a structure that can aid it in tracking a moving optimum, or indeed in quickly generating solutions when the optimum returns to previously examined neighbourhoods. In effect, a level of competency is developed within the population.

Finally, improving the evolvability of individuals was examined. Two approaches were analysed. The first focused on the genetic operators by explicitly identifying constructionally fit blocks of code and biasing reproduction to favour blocks that demonstrated a good statistical fitness in this area. This approach, however, was rooted in static environments and sought to encourage a quick convergence. Other studies focused on the issue of evolvability from a representational problem where the genotype and phenotype spaces are separated allowing many-to-one mappings and the evolution of neutral networks. Neutral networks were seen to improve the accessability of phenotypes where small changes in the genome could enable the discovery of newly fit phenotypes.

Moving on from approaches to aid algorithms, two related sections on performance measurement and benchmark problems were covered. These sections are related because the same criticism can be applied to both. Namely, if the goal of research into EC in dynamic environments is to bring the art closer to applicability to real-world problems, then the pursuit of good results for benchmarks and general performance measures is potentially a distraction. A danger exists that the race to deliver improvements to algorithms is being driven by one-up-manship on benchmarks and metrics, at the expense of understanding the nuances of real-world problems. A correlation between improvement on a benchmark problem and improvement on the real-world problem must be established if results on a benchmark are to carry any weight. This still considered, benchmark problems by their nature do not capture all the dynamics of complex real-world problems and should be considered more as unit tests or controls. In reviewing the categorisations in Table 3.2 the benchmark problems developed generally attempt to emulate two types of change as seen in Table 3.9. By extension, the metrics developed to evaluate the algorithms also focus on these two types. In attempting to follow through on the goals of research for EC in dynamic environments, it may instead prove more useful to research and code to the problems rather than benchmarks.

Table 3.9 The types of change faced in benchmarks, real-world problems, and nature

Type	Benchmarks	Real-World	Nature
Type		**Real-World Problems**	
Markov	X	X	X
Complex System	-	X	X
Deterministic	X	X	X

This then leads to a final issue: of all the literature covered in this chapter, not one study was conducted on the application of GP to dynamic environments. This is potentially due to a trend of working with benchmark problems for dynamic environments that focus on fixed-length individuals in bounded search spaces. The purpose of these benchmarks is to allow the results of different algorithms to be broadly comparable. However, the majority of the benchmarks are relatively simple and may potentially be holding back progression into the richer representation of GP.

3.7 Conclusion

This chapter has outlined the current state of the art in the area of EC applied to dynamic environments. Emanating from this process, first of all, is a clear categorisation of dynamic problems that exist, along with the types of changes that occur. This is achieved through the unification and development of different authors' works, which aided in the literature survey by presenting a clear perspective on what issues the various authors were seeking to address with their contributions.

Of these contributions, the incorporation of explicit memory comes at an efficiency cost while at the same time only providing a better performance for one type of change. The use of multiploid implicit memory also falls under the same criticism, but newer representations with genotype-to-phenotype mappings present promise for the efficient storage of previously useful genetic material. The development of building blocks within individuals and a competent population also serve as means to retrieve previously useful information without introducing computational costs.

The maintenance of a high level of dispersion and diversity is of great importance in developing evolutionary algorithms for dynamic environments. The implementation of mechanisms to allow this runs counter to the legacy development of the field, which has matured focusing on static problems where a speedy convergence is desired. Nevertheless a number of approaches have been described here with further opportunity presented in this study through the utilisation of GE's decoupling of search and solution spaces. Such a decoupling also provides scope for exploring evolvability along with taking into account the added flexibility through the incorporation of a BNF grammar.

There has been too much focus on the development of algorithms to narrow benchmarks, which has led the EC community to focus their efforts on bounded Markovian and Deterministic-type problems.

The most striking outcome of this survey of the literature on EC in dynamic environments is the lack of research in the application of GP to dynamic problems. In this book we endeavour to close that gap. This gap is addressed in a number of the experimental chapters beginning with more simple types of change and progressing to complex types of change in Chapter 8. In the next chapter we discuss Grammatical Evolution and its potential for application in these non-stationary domains.

Chapter 4
GE in Dynamic Environments

In the previous Chapter we highlighted the fact that very little research has been conducted into the area of Genetic Programming (GP) in dynamic environments. In this book we outline the foundations of research to date with Grammatical Evolution (GE) for these kinds of non-stationary environments. As described earlier, GE possesses a number of features that differentiate it significantly from GP and it is these features that present the most interesting avenues for exploration in relation to dynamic environments, more so than in their application to static problems.

In this chapter we start out by detailing in Section 4.1 the very first steps which we have taken with GE into the domain of non-stationary environments. Following this, in Section 4.2, we discuss the potential strengths of GE for the challenges presented by a dynamic world. Finally outline in Section 4.3 how we build the foundations upon which GE can be developed for application in these formidable environments.

4.1 The First Steps

In an investigation examining the possibility of evolving the grammar that GE uses [156] the Grammatical Evolution by Grammatical Evolution (GE^2) algorithm was detailed. This study focused on the utility of such an approach for dynamic symbolic regression problems where the target function was changed after a number of generations. The approach was initially inspired by an earlier study by Keller and Banzhaf [103], which examined whether or not it might be useful to evolve the genetic code for their linear form of GP. The net effect of their approach was to evolve biases for individual symbols of their programming language, including both functions and terminals. To this end the study was a success on the static problem examined.

By evolving the grammar that GE uses to specify a solution, one can effectively permit the evolution of the *genetic code*. The ability to evolve genetic code is important when one has little or no information about the

I. Dempsey et al.: Foundations in Gramm. Evol. for Dyn. Environ., SCI 194, pp. 55–68.
springerlink.com

problem being solved, or the environment in which a population exists is dynamic in nature and adaptability is essential for survival. A more adaptive and/or open-ended representation that can facilitate progression to different environments may be required to successfully tackle non-stationary domains. We now describe the $(GE)^2$ algorithm in detail and describe how it was applied to a dynamic environment.

4.1.1 Grammatical Evolution by Grammatical Evolution

When we have a set of production rules for a non-terminal, such as, for example, `<op>::= + | -`, a codon is used to select the rule to be applied during the development of a solution. In a similar manner to a biological genetic code, the productions above represent a degenerate genetic code by which a codon is mapped to a symbol in the output language [154]. A brief overview of the biological genetic code now follows.

In biology, a codon (on mRNA), which is comprised of a group of three nucleotides from the set $\{A, U, G, C\}$, is mapped to an amino acid from the set of 20 naturally occurring amino acids. In nature, the code is encoded in transfer RNA (tRNA) molecules, which have a domino like structure, in that one end matches (with a certain affinity dubbed the *wobble hypothesis*) to a codon, while the amino acid corresponding to this codon is bound to the other end of the tRNA molecule [118]. In this sense, the above productions are equivalent to two such tRNA molecules, one matching a set of codons to $+$ while the other matches a different set of codons to $-$. By modifying the grammar, we are changing the types of tRNA molecules in our system. To put it another way, we are directly modifying the genetic code by changing the mapping of codon values to different rules (amino acids).

In order to allow evolution of a grammar, $(GE)^2$, we must provide a grammar to specify the form a grammar can take. This is an example of the richness of the expressiveness of grammars that makes the GE approach so powerful. By allowing an EA to adapt its representation (in this case through the evolution of the genetic code or grammar) it provides the population with a potential mechanism to survive in dynamic environments. Such a representation also allows the automatic adaptation of biases during the search process.

In this approach we therefore have two distinct grammars, the *meta-grammar* (or grammars' grammar) and the *solution grammar*.[1]

[1] In the original study [156] we adopted the term *Universal Grammar* instead of meta-grammar. The notion of a universal grammar is adopted from linguistics and refers to a universal set of syntactic rules that hold for spoken languages [41]. It has been proposed that during a child's development the universal grammar undergoes modifications through learning that allows the development of communication in their parents' native language(s) [172]. We now prefer the use of the term meta-grammar as it is more firmly rooted in the Computer Science discipline of formal grammars.

In $(GE)^2$, the meta-grammar dictates the construction of the solution grammar. Given below are examples of these grammars for solutions that generate expressions, which could be used for symbolic regression type problems.

meta-Grammar

(Grammars' Grammar)

```
<g> ::=
      ''<expr> ::= <op> <expr> <expr> | <var>''
      ''<op> ::='' <ops>
      ''<var> ::='' <vars>

<ops> ::= <opt> ''|'' <ops>
      | <opt>

<opt> ::= + | - | * | /

<vars> ::= <vart> ''|'' <vars>
       | <vart>

<vart> ::= m | v | q | a
```

Solution Grammar

```
<expr> ::= <op> <expr> <expr>
         | <var>

<op> ::= ?

<var> ::= ?
```

In the example meta-grammar, a grammar, <g>, is specified such that it is possible for the non-terminals <var> and <op> to have one or more rules, with the potential for rule duplication. These are the rules that will be made available to an individual during mapping, and this effectively allows bias for symbols to be subjected to the processes of evolution. The productions <vars> and <ops> in the meta-grammar are strictly non-terminals, and do not appear in the solution grammar. Instead they are interim values used when producing the solution grammar for an individual.

The hard-coded aspect of the solution grammar can be seen in the example above with the rules for <op> and <var> as yet unspecified. In this case we have restricted evolution to occur only on the number of productions for <var> and <op>, although it would be possible to evolve the rules for <expr> and even for the entire grammar itself. It is this ability that sets this form of genetic code/grammar evolution apart from previous studies in GP. Notice that each individual has its own solution grammar.

In this study two separate, variable-length, genotypic binary chromosomes were used, the first chromosome to generate the solution grammar from the meta-grammar and the second chromosome the solution itself. Crossover operates between homologous chromosomes, that is, the meta-grammar chromosome from the first parent recombines with the meta-grammar chromosome from the second parent, with the same occurring for the solution chromosomes. In order for evolution to be successful it must co-evolve both the genetic code (otherwise known as the solution grammar) and the structure of solutions based on the evolved genetic code.

4.1.2 Experiments in GE^2 and Dynamic Environments

An instance of a symbolic regression problem was tackled in order to verify that it is possible for the co-evolution of a genetic code (or grammar) to occur along with a solution. A target function of $f(m,v,q,a) = a + a^2 + a^3 + a^4$ was chosen, with the three input variables m,v, and q introducing an element of noise. 100 randomly generated input vectors are created for each call to the target function, with values for each of the four input variables drawn from the range [0,1]. Runs were conducted with a population size of 100, for 100 generations. The other evolutionary parameters were as follows: pairwise tournament selection, generational replacement, bit mutation probability 0.01, one-point crossover probability 0.3, codon duplication probability 0.01. Wrapping is turned off, and codon lengths are initialised in the range [1,10], with a codon size of 8-bits. Fitness is minimisation of the sum of errors over the 100 test cases, and a protected division operator is adopted that returns one in the event of a division by zero. The progress of evolution toward the target solution can be seen in Fig. 4.1 with ever decreasing error at successive generations.

Fig. 4.1 shows the increasing frequency of occurrence of the target solution symbols a, $+$ and $*$. Curiously, after 50 generations the frequency of $*$ is dramatically less than a and $+$, and even less than $/$, even though there are double the number of multiplication symbols in the target solution as there are addition operators. It is not until after this point that we begin to see an increase in the frequency of $*$, which, although it finishes considerably lower than the other two symbols, finishes higher than all others. This could have implications as to how a solution to this problem is constructed, suggesting that firstly terms are added together with the use of multiplication not occurring until much later, perhaps replacing some of the addition operators, or secondly, through expansion of terms with the multiplication of a by itself.

The above results demonstrate that it is possible to co-evolve the solution grammar and solution specification with GE^2. Two experiments were then conducted where GE^2 was applied to two instances of dynamic symbolic regression.

4.1.3 Dynamic Symbolic Regression I

In addition to learning symbol bias, dynamic problems are another area in which one could expect to derive some benefit from using evolvable grammars by adapting these biases over time. In this case, one could reasonably expect a system with an evolvable grammar to be able to react more quickly to a change in the environment than a static one could, as a single change in a grammar can reintroduce lost genetic material. The target functions for the first instance are:

i. $f(m,v,q,a) = a + a^2 + a^3 + a^4$
ii. $f(m,v,q,a) = m + m^2 + m^3 + m^4$
iii. $f(m,v,q,a) = v + v^2 + v^3 + v^4$
iv. $f(m,v,q,a) = q + q^2 + q^3 + q^4$
v. $f(m,v,q,a) = a + a^2 + a^3 + a^4$

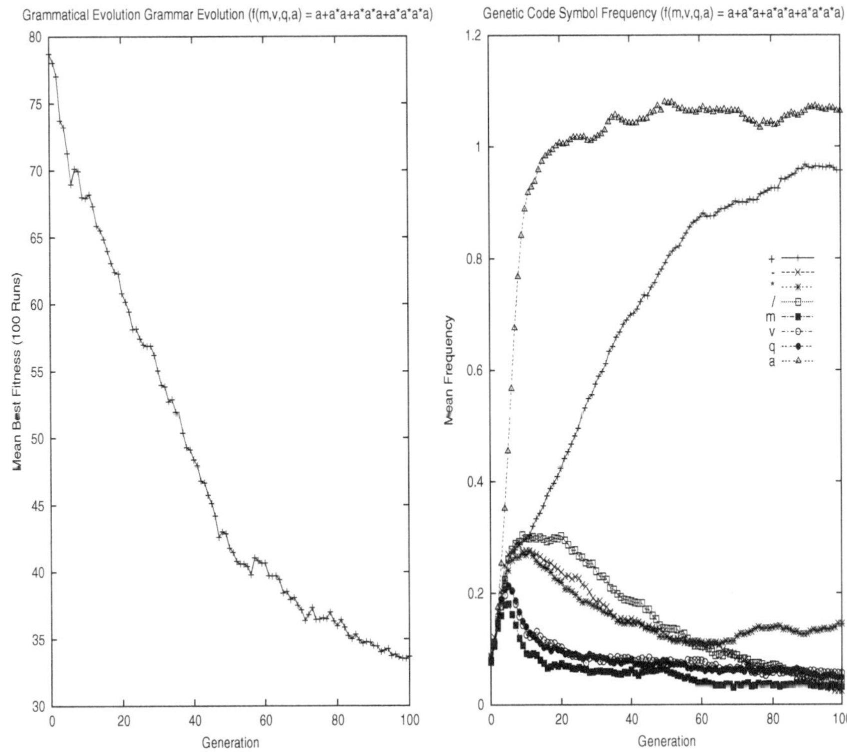

Fig. 4.1 A plot of the mean best fitness (left) and mean symbol frequency (right) from 100 runs of the quartic symbolic regression problem

The target changes between the functions above every 20 generations. The only difference between each successive function is the variable used. 100 randomly generated input vectors are created for each call to the target function, with values for each of the four input variables drawn from the range [0,1]. The symbols −, and / are not used in any of the target expressions. Runs were conducted with a population size of 500, for 100 generations, with all other parameters as reported earlier. A plot of the average best fitness and average symbol frequencies can be seen in Fig. 4.2. A sample of evolved grammars from one of the runs is given below, where in each case the grammar selected is the best solution from the generation just prior to a change in target.

Target 1

```
<op>::= +
<var>::= a
<expression>::= + a a
fitness: 34.6511
```

Target 2

```
<op>::= +
<var>::= m
<expression>::= + m m
fitness: 34.2854
```

Target 3

```
<op>::= + | -
<var>::= v
<expression>::= + v v
fitness: 36.6667
```

Target 4

```
<op>::= + | *
<var>::= q
<expression>::= + + q q * * q q * q q
fitness: 22.8506
```

Target 5

```
<op>::= + | *
<var>::= a
<expression>::= + * a + a a * a a
fitness: 7.85477
```

The results presented suggest that, when using dynamic grammars, it is possible to successfully preserve and improve solution structure, while still being able to learn appropriate terminal values. This is reflected in the fitness plot where, when the fitness function changes, in most cases there is a decrease in solution fitness for a short period when solutions adjust to the new variable

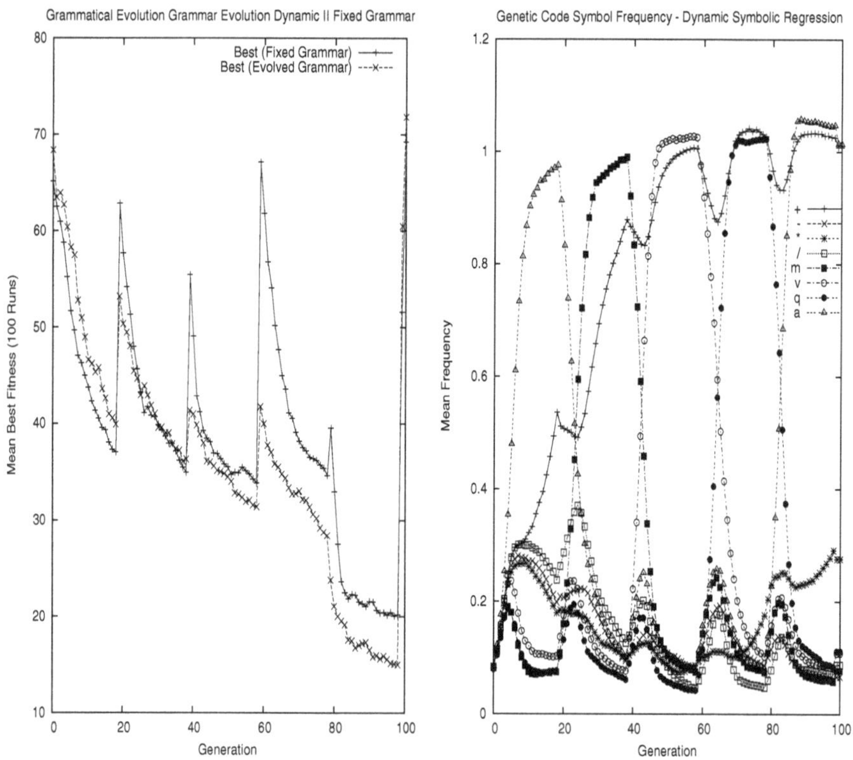

Fig. 4.2 Plot of the mean best fitness over 100 generations on the first dynamic symbolic regression instance with both static and dynamic grammars (left). Symbol frequency plot (right).

Table 4.1 Statistics for both the static and evolvable grammars on the first dynamic problem instance. Lower scores indicate better performance.

Fitness Case	mean fixed(dynamic)	median fixed(dynamic)	std. dev fixed(dynamic)	signif.
1	37.33 (40.55)	37.75 (38.22)	7.81 (10.082)	Yes
2	35.48 (36.08)	37.1 (36.57)	6.35 (8.73)	No
3	34.26 (31.53)	36.6 (36.48)	7.54 (10.79)	Yes
4	35.39 (28.74)	37.2 (35.08)	7.96 (12.46)	Yes
5	20.05 (15.1)	22.00 (20.54)	5.99 (10.17)	Yes

adopted. Later on in the simulations we reach the point where the structure becomes closer to the target and changes in variables alone no longer confer as much damage to fitness, which is again illustrated in the fitness plot (Figure 4.2).

A performance comparison of the dynamic and static equivalent of the grammar (given below) for this problem is presented in Fig. 4.2 and corresponding statistics can be found in Table 4.1.

```
<expr> ::= <op> <expr> <expr> | <var>

<op> ::= + | - | * | /

<var> ::= m | v | q | a
```

4.1.4 Dynamic Symbolic Regression II

The target functions for the second dynamic symbolic regression problem instance are:

i. $f(m, v, q, a) = a + a^2 + a^3 + a^4$
ii. $f(m, v, q, a) = m + a^2 + a^3 + a^4$
iii. $f(m, v, q, a) = m + m^2 + a^3 + a^4$
iv. $f(m, v, q, a) = m + m^2 + m^3 + a^4$
v. $f(m, v, q, a) = m + m^2 + m^3 + m^4$

The target changes between the functions above every 20 generations. The transition used in this problem differs from the previous in that only one term changes each time. However, the change is larger each time (because the power that the new term is raised to increases). 100 randomly generated input vectors are created for each call to the target function, with values for each of the four input variables drawn from the range [0,1]. The symbols q, v, $-$, and $/$ are not used in any of the target expressions. As in the previous dynamic symbolic regression problem instance runs are conducted with a population size of 500, for 100 generations, with all other parameters as per the standard values reported earlier. A plot of the average best fitness and average symbol frequencies can be seen in Figure 4.3.

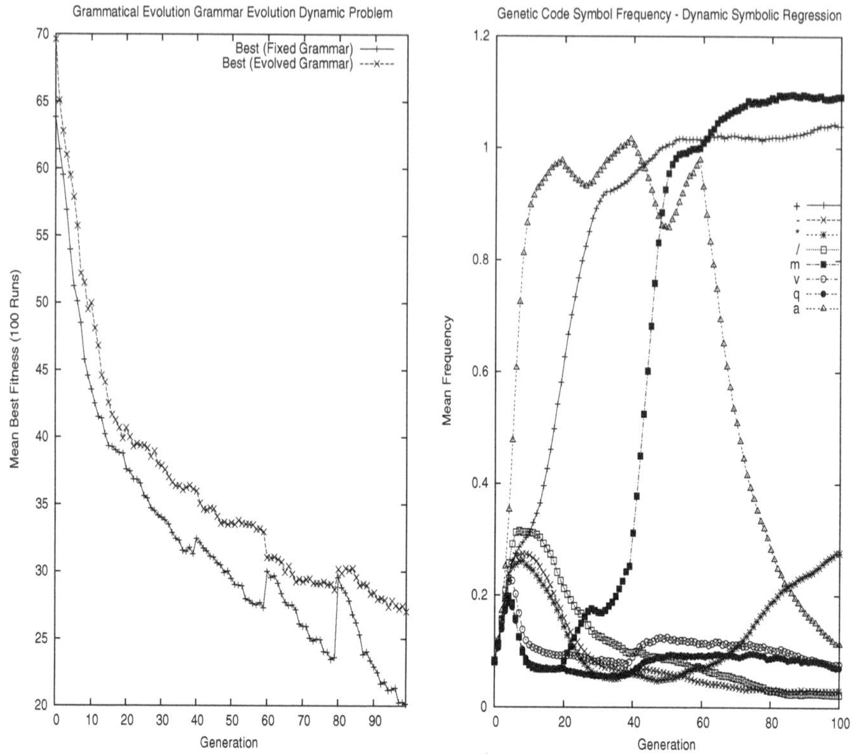

Fig. 4.3 Plot of the mean best fitness over 100 generations on the second dynamic symbolic regression instance with both dynamic and static grammars (left), and the mean symbol frequency (right)

It is interesting to note that fitness keeps improving over time for the evolvable grammar, with an occasional deterioration corresponding with a change in the fitness function. Notice how the disimprovement is more pronounced later in the runs, particularly for the static grammar, which is due to higher powers being exchanged. These results suggest that the evolvable grammar is more adaptable in scenarios with larger changes facilitating smoother transitions to successive targets. Also evident from Figure 4.3 is the manner in which the quantity of a in the population decreases over time while that of m increases. The two plots intersect at around generation 42, shortly after the target has changed to $f(m, v, q, a) = m + m^2 + a^3 + a^4$. However, the plots remain very close until around generation 60, at which time m^3 becomes part of the solution.

A sample of evolved grammars from one of the runs is given below, where the grammars presented represent the best solution at the generation just prior to each fitness change.

Target 1

```
<op>::= + | +
<var>::= a
<expression>::= (+ a a)
fitness: 37.4525
```

Target 2

```
<op>::= +
<var>::= m | a
<expression> = (+ a m)
fitness: 33.8423
```

Target 3

```
<op>::= + | *
<var>::= m | a
<expression> = (+ a (+ m (* a m)))
fitness: 22.9743
```

Target 4

```
<op>::= + | *
<var>::= m
<expression> = (+ m (* (+ (* m m) m ) m) )
fitness: 15.6311
```

Target 5

```
<op>::= + | *
<var>::= m
<expression>::= (+ (* (+ m (* m (+ m (* m m) m) m) ) ) )
fitness: 4.57967e-15
```

Table 4.2 Statistics for the second dynamic problem instance. Lower numbers indicate a better fitness.

Fitness Case	mean fixed(dynamic)	median fixed(dynamic)	std. dev fixed(dynamic)	signif.
1	39.27 (41.63)	37.98 (38.65)	9.18 (12.59)	No
2	31.55 (36.06)	31.93 (36.60)	6.77 (3.84)	Yes
3	27.62 (33.46)	25.82 (34.52)	6.3 (4.1)	Yes
4	24.05 (29.2)	22.62 (32.17)	5.83 (6.39)	Yes
5	21.34 (27.47)	18.74 (35.2)	11.42 (14.94)	Yes

A performance comparison of the dynamic and static equivalent of the grammar (static grammar as per earlier dynamic problem instance) for this problem is presented in Figure 4.3 and corresponding statistics can be found in Table 4.2. In this case the static grammar outperforms the evolving grammar in terms of best fitness values achieved for all targets but the first. With the evolving grammar there is, as usual, a warm up period where a suitable grammar must be adopted before good solutions can be found. When successive targets are very similar to previous ones this almost negates the potential benefits that a more adaptive representation can bring, as in the case of the evolvable grammars. Clearly, some dynamic problems are more dynamic than others as discussed in Chapter 3, especially in terms of the degree of change. Previous work (e.g., [144]) with GAs applied to dynamic problems has shown that, when the change is relatively small, a standard GA with high mutation can handle those types of problems. We believe it is likely to be the same for GP.

These results would also lend support to the idea of introducing different operator rates on the grammar chromosome to the solution chromosome, allowing the population to converge towards a similar grammar, facilitating the exploration of solutions based on a similar grammar. If these rates were adaptable, then it may be possible to allow grammars to change more often if the target changes are large, and vice versa.

This first study in the application of GE to dynamic environments was encouraging as it suggested that there may be benefits to representational flexibility, which can be provided with a GE approach. The following section describes some of the potential strengths of GE for dynamic problems.

4.2 Strengths

In [148] O'Neill described eight desireable features, inspired by Molecular Biology, that evolutionary algorithms could incorporate as researchers attempt to improve their algorithms' performance. Seven of these, implemented in GE, are as follows:

i. **Generalised encoding that can represent a variety of structures**
Through the decoupling of search and solution spaces, a mapping process can afford the opportunity to generate phenotypes in an arbitrary language.

ii. **Efficiency gains for evolutionary search**
Improvement of the evolutionary search is generally positive. By incorporating degenerate genetic code, a medium is provided for neutral mutations to occur giving rise to neutral evolution. As described in Sections 3.3.5 and 3.3.6, this yields efficiency gains in evolution.

iii. **Maintenance of genetic diversity within an evolving population**
Diversity was highlighted as being one of the key approaches for evolutionary algorithms in dynamic environments in Chapter 3. An ability to maintain a diverse dispersed population allows the algorithm to conduct wide coverage of the search space, equipping it with the ability to quickly discover new optima when the environment changes.

iv. **Preservation of functionality while allowing continuation of search** Providing a mechanism to conduct neutral mutations allows an algorithm to preserve a functioning, fit, phenotype while continuing the evolutionary search. This feature gives rise to the potential of Engelhart's neutral networks being evolved. Individuals aligned along neutral networks in the genotypic space may then easily evolve to produce alternate phenotypes.

v. **Reuse of genetic material**
Reusing genetic material presents efficiency gains in an implemented algorithm. When individuals in a population are of variable length, the possibility for *bloat* arises whereby the amount of genetic material in the individuals may grow over the course of the evolutionary process.

vi. **A compression of representation**
Again regarding an implemented algorithm this feature bears benefits at execution.
vii. **Positional independence**
By decoupling the expressed functionality of a gene from its position on the chromosome a mechanism is provided to preserve the functionality of genes after crossover. This may lead to more productive recombination events.

Standard GE, as described in [148], integrated the first six features with further research leading to the implementation of the seventh feature in πGE [157]. Considering the various approaches for evolution in dynamic environments identified in Section 3.3, some of these features bear special relevance when applied to dynamic environments.

The first feature of adopting a generalised encoding is achieved in GE through the use of the BNF grammar plug-in. With regard to dynamic environments, the development of the grammar is of particular importance as it affords the modeller the opportunity to incorporate some fundamental domain knowledge into the system while maintaining flexibility. Recent developments in GE have lead to the development of meta-Grammars and Grammatical Evolution by Grammatical Evolution. In this case a meta-Grammar or grammar's grammar is defined, which allows evolution to evolve its own vocabulary for the expression of phenotypes. Chapter 6 will deal with this in more detail.

Regarding efficiency gains for evolutionary search, O'Neill refers to the presence of degenerate genetic code and its relationship with Kimura's neutral theory of evolution [108]. The degenerate or redundant genetic material is brought about in GE through the many-to-one mapping from genotype to phenotype, allowing neutral mutations to occur. This caters directly to dynamic environments as it allows for the development Engelhart and Newman's neutral networks [140] as already outlined in Section 3.3.5 and provides a representation that is more evolvable than a direct encoding as outlined in Section 3.3.6.

With the employment of the genotype-to-phenotype mapping and the presence of degenerate genetic code, features three and four are simultaneously addressed. Due to the unconstrained search in the genotype space with a many-to-one mapping back to the phenotype solution space, there is the potential for a high degree of diversity to be maintained continuously over the life cycle of the algorithm. Whether this is in a static or dynamic environment, it places this feature in the category of preemptive approaches used to maintain diversity.

The possibility of neutral evolution brought about by degenerate genetic code is also responsible for a robustness in the solution space to changes in the genetic search space. This is because as mutations can occur with the presence of degenerate code and still produce the same phenotype. This mechanism then allows GE to preserve the expressed functionality of the phenotype while continuing with a neutral evolutionary search in the genotypic space dispersing across a neutral network.

Features five and six are implemented in GE through the use of the wrapping operator. In other words, if upon reaching the end of the genome, an individual is still not a fully formed phenotype, mapping will continue again at the beginning of the genetic string. Though O'Neill found the use of this operator to provide a better success rate than with the operator turned off, the operator does not address directly any of the extra issues involved in dynamic environments. Table 4.3 presents a summary of the implemented features in GE.

Table 4.3 Implementation of features in GE

	Feature	**Implementation**
1	Generalised Encoding	BNF Grammar
2	Efficiency Gains	Degenerate Genetic Code
3	Diversity	Many to One Mapping
4	Preservation of Functionality During Search	Many to One Mapping
5	Re-use of Genetic Code	Wrapping Operator
6	Compression of Representation	Wrapping Operator
7	Positional Independence	πGE

4.3 Extending GE for Dynamic Environments

Aside from [156], no work has been done in exploring the use, and analysing the performance, of GE in dynamic environments prior to the research outlined in the remainder of this book. In light of this, shortfalls specific to GE's proficiency in these environments have not been identified. Considering that GE has undergone little analysis in dynamic environments, a number of the features built-in to the design of the paradigm bear particular significance when placed in a dynamic environment. The addition of BNF grammars as an input to GE present an extra level of representational flexibility that may be exploited in terms of evolvability. With BNF grammars, the researcher may incorporate a greater level of adaptability in the vocabulary available to the search process while ensuring syntactic correctness. Through the design of the grammar, a structure may be created that aids in the evolution towards fitter solutions. The ability to allow the evolutionary process to evolve its own grammars takes this feature even further. Reflecting upon the approaches highlighted in this book so far, with the inclusion of degenerate genetic code and many-to-one mapping, standard GE is equipped with an ability to form neutral networks in the genotypic space through neutral evolution and to preemptively maintain a high degree of dispersed diversity unconstrained by a phenotypic search-space boundary.

In placing the features of GE in the context of tables developed in this chapter, the inclusion of the BNF grammar allows researchers to scope the adaptiveness of the vocabulary. It also presents an opportunity to incorporate useful domain knowledge tailoring it for problems where information can be extracted out of the domain. Added to this, the BNF grammar can aid in the evolvability of individuals thus aiding in the evolution for all types of change. Finally, the maintenance of diversity through many-to-one mapping is a feature that enables wide coverage of the solution space as evolution progresses. While all problems benefit from this coverage of the search space, the maintenance of diversity throughout the life cycle of a run is tailored for Markov and Complex type changes. Though Deterministic problems will also benefit from this diversity early in the search, a level of convergence is sought in uncovering the predictable nature of the change. Therefore the maintenance of diversity is of lesser importance for this type of problem. Table 4.4 summarises these features.

Table 4.4 Features of GE that cater to the types of change faced in dynamic environments

Type	BNF Input (Evolvability)	Many-to-One (Diversity)
Markov	X	X
Complex System	X	X
Deterministic	X	-

A problem identified in GP in general, and also existing in GE, is its inflexible approach to the creation and variance of constants [9]. GE's approach to the variance of constants is through the evolution of expressions operating on a handful of constants that are defined in the BNF grammar prior to the system's execution. This weakness presents particular problems when conducting evolution in a dynamic environment, as for both efficiency and flexibility purposes, it may be desireable to directly adapt old or evolve new constants in the solutions as the optimum shifts or moves on the landscape. Indeed, in many of these problems a continual adaptation of constants is required and improvements in this area will be useful in applying GP and GE to dynamic problems. This weakness will be addressed in Chapter 5. It is also worth noting that even though [156] did apply $(GE)^2$ to dynamic environments it did so with very little analysis pertaining to the dynamic domain. Chapter 6 will endeavour to take this analysis further and examine the effects of the extra dimension of adaptability in the algorithm.

Through examining the extra layer of adaptability in $(GE)^2$ and adapting a population of solutions as time progresses the representation of solutions as well as the solutions themselves are being adapted. Another facet of GE may also be adapted with the progression of time–dynamic adjustment of the parameters of the algorithm. These include, for example, the number of generations of evolution and probabilities for applying genetic operators. Work has been conducted on investigating the effects of adapting these parameters for static environments [141].

4.4 Conclusion

In conjunction with the review of EC for dynamic environments and our knowledge of GE to date we have identified a number of research gaps. These are summarised as follows:

- **Analysis of the level of diversity maintained in GE promoted by the separation of search and solution spaces**
  The maintenance of a dispersed, diverse population is of great importance in dynamic environments; the separation of search and solution spaces may aid this. An examination of this is undertaken in Chapter 8.
- **Exploiting the use of BNF grammars to improve evolvability**
  Not only does GE provide for improved evolvability through potential search efficiencies presented due to its genotype-to-phenotype mapping and subsequent separation of its search and solution spaces. In addition, the opportunity exists to improve evolvability through the use of the BNF grammar used by GE. The majority of the research presented in this book focuses on this form of representational evolvability. Chapters 5 and 6 explore this.
- **Use of a real-world problem as a benchmark, and the application of problem domain specific analytics to determine behaviour of algorithm developments**
  The use of such a benchmark and analytics provide both qualitative and quantitative evidence of algorithm performance and also facilitates comparison with human achievements. Chapter 8 conducts such analysis in the financial domain of trading.
- **Application of GP type paradigm to a dynamic problem**
  The lack of research in the application of GP to dynamic problems is remarkable, this book will endeavour to close that gap. This gap is addressed in a number of the experimental chapters beginning with more simple types of change and progressing to complex types of change in Chapter 8.

In subsequent chapters these gaps shall be examined in greater detail.

Chapter 5
Constant Creation and Adaptation in Grammatical Evolution

Fundamental to many applications of Genetic Programming [112, 113, 22, 152] is the generation and use of constants. Hence the discovery of an efficient means for creating diverse constants is important. When the domain in which a GP system operates is dynamic, an even greater level of adaptability is required as it may become necessary to generate novel constant values with changes in the environment.

This chapter introduces and explores two novel constant-generation schemes; *Digit Concatenation* and *Persistent Random Constants*, as a means of generating constants in GE. The impetus for the introduction of these schemes is to address the existing weaknesses in the state of the art in GP for the generation and adaptation of constants, and to explore the potential of alternative representations in adaptation. While current schemes perform satisfactorily in static scenarios where a level of convergence is required, by migrating them to dynamic environments, limitations are encountered. This chapter explores the different constant creation methods with the aim of identifying a flexible and adaptive constant creation approach that will satisfy the extra requirements of evolving in a dynamic environment.

Following an introduction, Section 5.2 outlines existing techniques used in Genetic Programming to create constants. Section 5.3 examines the performance of the Digit Concatenation method in comparison with the traditional method for generating constants within GE. Section 5.4 compares the Digit Concatenation method with another novel method of constant creation based on Ephemeral Random Constants (ERC). Section 5.5 takes this comparison further by providing the Digit Concatenation method with the ability to evolve expressions. An analysis of constant representation and its impact on problem difficulty is detailed in Section 5.6. Finally Section 5.7 arrives at conclusions as to which method of constant generation is best.

5.1 Introduction

Integral to the proper functioning of most GP solutions is the incorporation of constants: numbers are passed as parameters to functions, to scale return

I. Dempsey et al.: Foundations in Gramm. Evol. for Dyn. Environ., SCI 194, pp. 69–104.
springerlink.com

values, as constants in expressions and in many other roles. Evolutionary Algorithms (such as GAs and Evolutionary Strategies) have been shown to be successful in evolving specific constants as parameters to functions. However, in GP, constants are generally co-evolved in line with the structure and other program primitives. Little research has been undertaken to address this area. Specifically, Koza states that:

"... the finding of numeric constants is a skeleton in the GP closet... [and an] area of research that requires more investigation..." [63]

The issue Koza refers to is that constants in GP are represented as terminal nodes on the GP tree structure and, therefore, are unaffected by the canonical GP operators of crossover and mutation. This is because these are applied only to the tree structure itself. The result then is that new values are only arrived at through expressions between these nodes. The problem is magnified by the fact that constants in GP can be evolved out of the population as evolution progresses, leading to a reduction in the diversity of contants available to individuals. A special form of point mutation must be introduced to allow the values of the constants in any one node of a tree to be modified. Otherwise the generation of new constants can only be achieved through the recombination of existing constants in expressions.

In dynamic problems, the issue becomes more acute as the environment may shift significantly over time. This necessitates a greater level of adaptability among the constants or indeed the generation of new constants. Considering this, a number of requirements emerge as to what would be desirable in a constant generation mechanism for dynamic environments.

i. **Static Targets:** The mechanism naturally should be able to quickly and efficiently evolve solutions for static problems. Evidence of this will serve as a control and demonstrate that the mechanism works.
ii. **Types of Change:** All the types of change described in Chapter 3 can require the generation and adaptation of constants. Therefore the new mechanism should be able to perform under these circumstances. Reflecting upon the approaches outlined in Chapter 3, a potential solution may be able to maintain a dispersed level of diversity in order to achieve or allow a level of problem decomposition to form building blocks of useful relevant constants.
iii. **Adaptable:** The constants themselves should be adaptable so as to permit evolution to alter expressions by small degrees. To enable this, a constant generation mechanism must allow the actual numbers to be evolved themselves so as to conduct local searches of promising constants. This also ties in with Evolvability as outlined in Section 3.3.6. Such a mechanism should exhibit a high level of *latent evolvability* identified by Reisinger [179].
iv. **New Constants:** The ability to generate new constants is imperative in dynamic environments. The presence of such a feature will allow evolution to explore new areas of the solution landscape and also maintain diversity.

v. **Large and Small Constants:** A mechanism for generating constants for an evolutionary system operating in a dynamic environment should be equipped with the flexibility to generate numbers of any size – not just within a fixed range, decided upon a-priori.

This chapter will aim to highlight limitations in existing approaches to constant creation and adaptation. Two novel constant generation methods will be introduced which aim to address the above requirements for constant generation mechanisms for dynamic environments. While this chapter does provide coverage over different types of dynamic environments, it is not exhaustive and experimentation in later chapters will broaden the scope to simulate more complex types of change and different deterministic types. Over the course of the experiments undertaken, the most appropriate form of constant generation will be identified.

5.2 Constant Generation in GP

Ephemeral Random Constants are the standard approach to constant creation in GP, having values created randomly within a pre-specified range at the initialisation of a run [110]. The terminal $\Re$ is specified in the GP terminal set. At generation zero, whenever this symbol is encountered at a leaf node on the tree representation it is replaced with a random number generated within the pre-specified range. These values are then fixed throughout a run, and new values can only be arrived at through combinations of these values and other items from the function and terminal set in expressions.

A number of variations on the ephemeral random constant concept have been applied in tree-based GP systems, all of which have the common aim of making small changes to the initial constant values.

Constant perturbation [207] allows GP to fine-tune floating point constants by rescaling them by a factor between 0.9 and 1.1. This has the effect of modifying a constant's value by up to 10% of its original value.

Numerical terminals and **numerical terminal mutation** were used in [5]. The numerical terminal mutation operator selects a real valued numerical terminal in an individual and adds a Gaussian distributed noise factor, such that small changes are made to the constant values.

The **numeric mutation** operator [63] replaces the numeric constants in an individual with new ones drawn at random from a uniform distribution with a pre-specified range. The selection range for each constant is specified as the old value of that constant plus or minus a temperature factor.

Linear scaling [96, 143, 102] has been used to optimise values within their local neighbourhood. It is performed using linear regression on the values expressed, where a line is derived to fit the data and new values are explored in the neighbourhood.

A study in [194] used two forms of constant mutation, **creep** and **uniform** mutation, where values are altered by a small amount or mutated to a randomly different number. The study found greater benefits in uniform mutation where the ability to introduce new constants into a population as evolution progresses and maintain a highly diverse array of constants is generally beneficial to the fitness of individuals.

Banzhaf [9] notes the *"inflexible approach"* adopted in Koza's ERC and uses a genotype-to-phenotype mapping to interpret 10 bits as a natural number in the range 0 to $2^{10} - 1$. Then, depending on the prior symbol being consumed, this random number is mapped to an interval of real or integer constants. The benefit of this approach is that now standard mutation and crossover operators can be applied directly to the genotype, which forms the numerical phenotype. Effectively, the constants become parameters that are coded on the genotype.

With the exception of the final two approaches, each of these methods uses Ephemeral Random Constants as its base with the focus on changing the original random values by small amounts to improve fitness. None of these approaches, however, address a fundamental flaw in ERC. The flaw is that once a constant is evolved out of the population of solutions, it cannot be re-introduced to the population at a later stage. The phenotypic representation of GP does not facilitate a mechanism whereby formerly successful constants can be recalled. This leads to an inevitable decline in the diversity of constants that are available for solutions to exploit, unless a genetic operator is introduced to address this specific problem. All of these approaches also suffer from the weakness that the constants are generated within a certain range, which is built into the system prior to its execution. Daida et al., have demonstrated that the standard ERC approach to constants can escalate problem difficulty on certain problems when the range of ERC's provided increases [48]. This can then present issues where the problem is dynamic in nature.

GE can borrow from the experience of GP by extending the established methodology and introducing a new form of constant creation that potentially addresses the issue of beginning an evolutionary run with a fixed range of constants and offers the possibility of creating new values over the course of a run. With this in mind, the utility of this novel approach is determined by examining it under GE's capacity to create and adapt constants in isolation in order to gather a clear view of its behaviour and relative performance.

The experiments in this chapter are focused on identifying the best method for constant creation and adaptation with a emphasis on dynamic environments. In doing this,r the opportunity is also presented to examine the behaviour of the evolutionary process, and GE itself, in dynamic environments. This is useful as the types of dynamic problems tackled here present less complex types of change and so make the analysis of GE's behaviour more amenable.

5.3 Evolving Constants Using Digit Concatenation

We now introduce constant creation in GE illustrating the standard approach before detailing the new digit concatenation representation. The objective of this section is to determine whether Digit Concatenation can outperform the traditional expression-based approach to constant creation in GE.

5.3.1 Traditional Constant Creation in GE

The traditional approach to constant generation in GE relies upon defining a handful of constants in the BNF grammar, with the recombination of these terminals using expressions and function terminals leading to the creation of "new" values. Below is an example of a grammar that adopts such an approach.

```
<value> ::= <value> <op> <value>
          | ( <value> )
          | <number>
          | <func> ( <number> )
<func> ::= sin | cos | tan
<op> ::= + | - | / | *
<number> ::= 0 | 1 | 2 | 3 | 4 | 5 | 6 | 7 | 8 | 9
```

Here the grammar is provided with ten constants, operators, and trigonometric functions as terminals. The grammar may then combine these terminals to form expressions using the first production rule, with a sample output looking like the following:

```
4 + 7 * (sin ( 8 + 7 ) )
```

5.3.2 Digit Concatenation in GE

The Digit Concatenation method for constant creation provides GE with the fundamental building blocks for the construction of numerical values. An example of a grammar using Digit Concatenation is given below.

```
<real> ::= <int><dot><int> | <int>
<int> ::= <int><digit> | <digit>
<digit> ::= 0 | 1 | 2 | 3 | 4 | 5 | 6 | 7 | 8 | 9
<dot> ::= .
```

which could produce a sample output of

```
20034.51
```

In this grammar the digits zero to nine are used as the fundamental building blocks to create all other numbers simply by concatenating them together. By providing GE with the ability to create and evolve constants in this fashion, a number of the shortcomings of ERC in GP are potentially addressed.

- The problem of constant creation is approached from a fundamental representational level by addressing how evolution can create and adapt constants rather than providing it with a fixed pool.
- The requirement in ERC of generating the initial pool of numbers within a pre-specified range is overcome. Through using Digit Concatenation, constants of any size or floating-point precision can be created with evolution determining their usefulness.
- The constants themselves can be adapted and evolved in order to conduct local searches around relatively fit phenotypes.
- The facility exists to re-evolve previously successful constants.

Such features can then provide the ability for GE to maintain a satisfactory level of diversity of constants and allow the formation of useful strings of constants that can spread through the population, linking back to the approaches outlined in Chapter 3.

5.3.3 Problem Domain and Experimental Approach

Two grammars were constructed, one using Digit Concatenation and the other incorporating the Traditional method. The aim of these experiments is to examine the performance of GE in the creation and adaptation of constants in isolation, without involving any of the complexity of evolving the constants in line in programs. This is left to subsequent chapters. In this manner the experiments focus on the issue in its own right before adding context. Considering the requirements laid out in Section 5.1 and the types of change identified in Chapter 3, the performance of these grammars is measured on three different types of constant creation problems. The constant creation problems include: finding a static real constant in order to determine if the mechanism is functional under static conditions; satisfying the first requirement in Section 5.1; and finding dynamic constants where Markov and Deterministic types of change are tackled. In this manner these problems also represent a simplified testbed for the analysis of GE in dynamic environments while at the same time provding insights on how to create and evolve constants.

Finding a Static Real Constant

The aim of this problem is to evolve a single real constant and present a vanilla-type static problem to the mechanisms under examination. Three target constants of increasing difficulty were selected arbitrarily: 5.67, 24.35, and 20021.11501. Fitness is defined as the absolute difference between the target and the evolved values; the goal being to minimise the difference.

Finding Dynamic Real Constants

These experiments examine problems under two categories of change. The first test presents a Markov type problem. The target shifts through a sequence to a different real value every 10 generations. The targets in this experiment are a random sequence of values: 24.35, 5.67, 5.68, 28.68, and 24.35. For the next experiment, using the same fitness function, a second set of targets with a Deterministic oscillatory type of change is presented where the targets alternate between 24.35 and 5.67 every 10 generations. The aim of these problems is to make it easier to compare the different constant generation methods in terms of their ability to adapt to a changing environment. Further, the problems investigate their behaviour in the event of changes on both a small and a large scale. As in the static problem, fitness in this case is the absolute difference between the target and evolved values, with the goal being the minimisation of this difference.

The Logistic Difference Equation

A more complex type of regression problem is also tackled, where GE is tasked with evolving a coefficient for the logistic difference equation which exhibits chaotic behaviour. With systems exhibiting this behaviour, long-term prediction is problematic as even a small error in estimating the current state of the system leads to divergent system paths over time. Short-term prediction however, may be feasible [91]. Because chaotic systems provide a challenging environment for prediction, they have regularly been used as a test bed for comparative studies of different predictive methodologies [142, 38, 195]. In this time series, information is drawn from a simple quadratic equation, the logistic difference equation.

$$x_{t+1} = \alpha x_t(1 - x_t) \qquad x \in (0.0, 1.0) \tag{5.1}$$

The behaviour of this equation is crucially driven by the parameter α. The system has a single, stable fixed point (at $x = (\alpha - 1)/\alpha$) for $\alpha < 3.0$ [195]. For $\alpha \in (3.0, \approx 3.57)$ there is successive period doubling, leading to chaotic behaviour for $\alpha \in (\approx 3.57, 4.0)$. Within this region, the time series generated by the equation displays a variety of periodicities, ranging from short to long [129]. In this study, three time series are generated for differing values of α. The choice of these values is guided by [129], where it was shown that the behaviour of the logistic difference equation is qualitatively different in three regions of the range (3.57 to 4.0). To avoid any bias which could otherwise arise, parameter values drawn from each of these ranges are used to test the constant evolution grammars. The goal in this problem is to rediscover the original α value. As this equation exhibits chaotic behaviour, small errors in the predicted values for α will exhibit increasingly greater errors from the target behaviour of this equation with each subsequent time step. Fitness in this case is the mean squared error, which is to be minimised. 100 initial values for x_t were used in fitness evaluation, and for each x_t iterating 100 times (i.e. x_t to x_{t+100}). Because of this, there will be a strong feedback from the problem back to GE with slightly more accurate solutions achieving a much stronger fitness and so standing a stronger chance of being selected for reproduction. The goal of this experiment is to evolve closely accurate individuals.

Constant Creation Grammars

The grammars adopted are given below. The Digit Concatenation grammar (Cat) only allows the creation of constants through the concatenation of digits, whereas the Traditional grammar (Trad) restricts constant creation to the generation of values from expressions. The Traditional grammar is only provided with the basic mathematical operators to simplify the analysis of its behaviour.

Digit Concatenation (Cat) Grammar

```
<value> ::= <cat>
<cat> ::= <int><dot><int> | <int>
<int> ::= <int><number> | <number>
<number> ::= 0 | 1 | 2 | 3 | 4 | 5 | 6 | 7 | 8 | 9
<dot> ::= .
```

Traditional (Trad) Grammar

```
<value> ::= <value> <op> <value>
          | ( <value> <op> <value>)
          | <number>
<op> ::= + | - | / | *
<number> ::= 0 | 1 | 2 | 3 | 4 | 5 | 6 | 7 | 8 | 9
```

5.3.4 Results

For each grammar on every problem instance, 30 runs were conducted using population sizes of 500, running for 50 generations on the static and dynamic constant problems in order to examine the efficiency of the methods. It also included 100 generations for the logistic difference equation, adopting one-point crossover at a probability of 0.9 and bit mutation at 0.1. Roulette selection and a generational rank replacement strategy of 25% was adopted where the weakest performers were replaced by the newly generated offspring. The GE wrapping operator was allowed to perform 8 wrapping operations before terminating and a binary 8-bit encoding was used for the individuals. A random initialisation was conducted up to 104 bits long with evolution allowing this to vary after the first generation. A section describing the results of each of the experiments is provided next, followed by a discussion of their implications.

Finding a Static Real Constant

On all three instances of this problem, a t-test and bootstrap t-test [26] (5% level) on the best fitness values reveal that the Digit Concatenation grammar significantly outperforms the standard expression-based approach. Performance statistics for each grammar are given in Table 5.1, and a plot of the mean best fitness at each generation for the three targets can be seen in Figure 5.1.

Table 5.1 Statistics for the best fitness values (lower value is better) at generation 50 on the static real constant problem

Target Constant	Grammar	Mean	Median	Std. Dev.
5.67	Trad	0.33	0.33	0.0
	Cat	0.0	0.0	0.0
24.35	Trad	0.36	0.35	0.055
	Cat	0.002	0.0	0.009
20021.11501	Trad	7741.35	10000	3828.9
	Cat	1005.24	0.91	3049.5

Notably, the Trad grammar did not perform as well as the Cat grammar in evolving the large number by a significant margin. This demonstrates that a grammar with the Digit Concatenation approach to constant creation is significantly better at generating larger numbers. It is worth stressing that larger numbers could just as easily be large whole numbers or numbers with a high degree of precision (real).

Finding Dynamic Real Constants

For the first instance of this problem where the successive target constant values are 24.35, 5.67, 5.68, 28.68, and 24.35 over the course of 50 generations, performance statistics are given in Table 5.2, and a plot of mean best fitness values for each grammar can be seen in Figure 5.2.

Performing a t-test and a bootstrap t-test on the best fitness values at generations 10, 20, 30, 40, and 50, it is shown that there is a significant (5% level) performance advantage in favour of the Digit Concatenation grammar up to generation 30. However, beyond this point, the advantages of one grammar over the other are not as clear cut.

In the second instance of this problem, where the target constant values oscillates every 10 generations, between 24.35 and 5.67 over the 50 generations, again a similar trend is seen. In this case, the Digit Concatenation grammar is significantly better at the 5% level than the Traditional grammar at each of the 10, 20, 30, 40, and 50 generations. However, this difference is decreasing

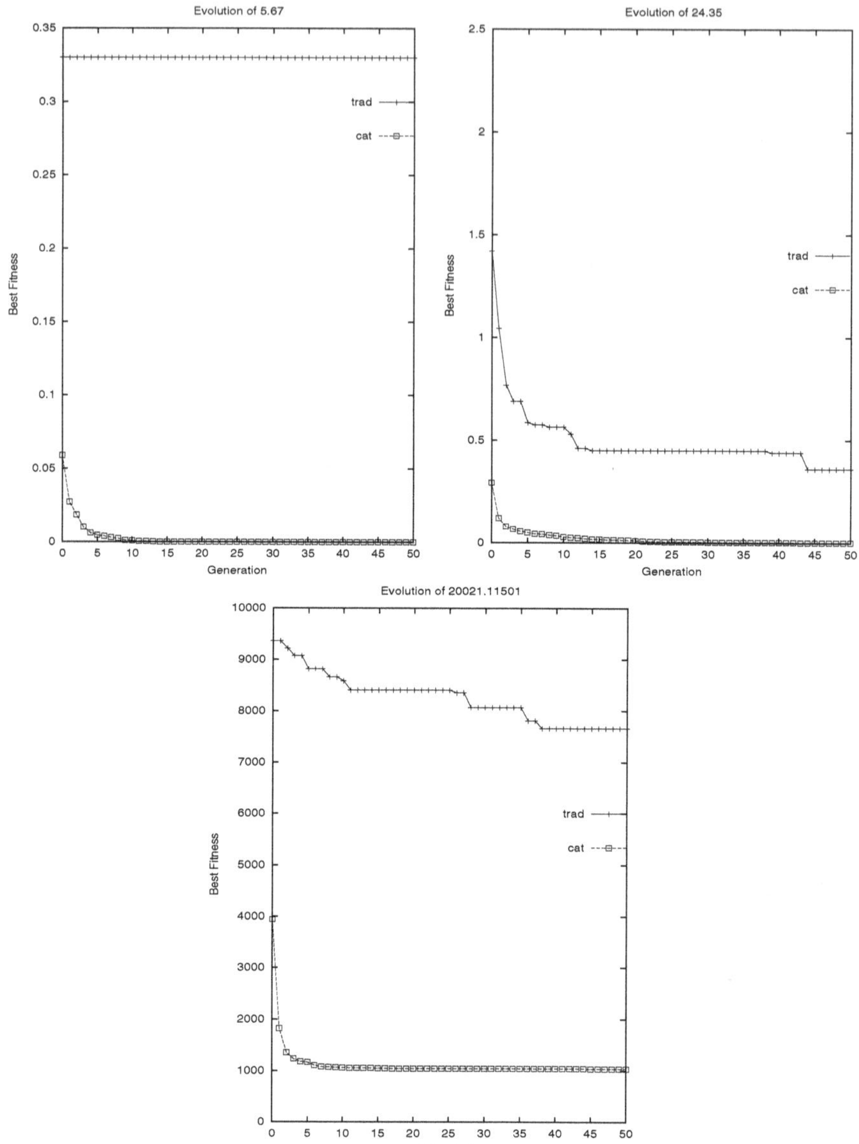

Fig. 5.1 Mean best fitness values (lower values are better) plotted against generations for each of the four grammars. Target values are 5.67 (top left), 24.35 (top right), and 20021.11501 (bottom).

over time. From the results of both of these dynamic problem instances, there are clearly adaptive advantages to using the Digit Concatenation grammar over the traditional expression-based approach.

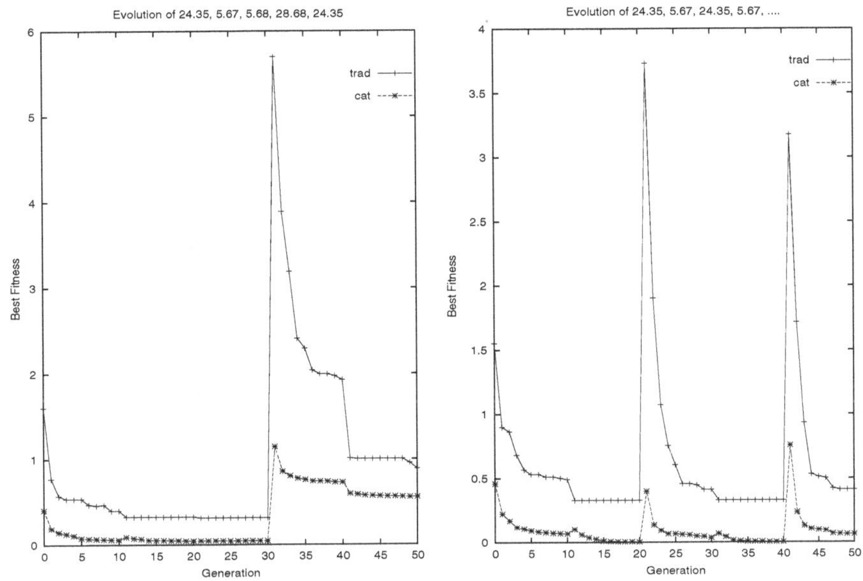

Fig. 5.2 Mean best fitness values (lower values are better) plotted against generations for each of the three grammars. Target values are 24.35, 5.67, 5.68, 28.24.35 (left) and 24.35, 5.67,... (right).

Table 5.2 Statistics for the best fitness values (lower value is better) in the dynamic real constant problem

Generation	Target Constant	Grammar	Mean	Median	Std. Dev.
10	**24.35**	Trad	0.4	0.35	0.114
		Cat	0.061	0.01	0.133
20	**5.67**	Trad	0.33	0.33	0.0
		Cat	0.047	0.0	0.17
30	**5.68**	Trad	0.32	0.32	1.129e-16
		Cat	0.046	0.0	1.724e-01
40	**28.68**	Trad	2.063	1.5	3.474
		Cat	0.046	0.0	1.724e-01
50	**24.35**	Trad	0.937	0.35	2.755
		Cat	0.541	0.002	2.799

The Logistic Difference Equation

The results for all three instances of this problem can be seen in Table 5.4 and Figure 5.3. Statistical analysis using a t-test and bootstrap t-test (5% level)

Table 5.3 Statistics for the best fitness values (lower value is better) in the oscillating dynamic real constant problem

Generation	Target Constant	Grammar	Mean	Median	Std. Dev.
10	**24.35**	Trad	0.507	0.35	0.426
		Cat	0.089	0.011	0.193
20	**5.67**	Trad	0.33	0.33	0.0
		Cat	0.005	0.0	0.0167
30	**24.35**	Trad	0.487	0.35	0.426
		Cat	0.046	0.022	0.07
40	**5.67**	Trad	0.33	0.33	0.0
		Cat	0.0004	0.0	0.01
50	**24.35**	Trad	0.487	0.35	0.426
		Cat	0.061	0.014	0.131

reveal that the Digit Concatenation grammar significantly outperforms the traditional constant creation approach on each problem instance, successfully rediscovering the target α in each case.

Table 5.4 Statistics for the best fitness values (lower value is better) in the logistic difference equation problem

Target Constant	Grammar	Mean	Median	Std. Dev.
3.59	Trad	6.074e-03	6.074e-03	2.647e-18
	Cat	4.818e-07	3.902e-19	1.249e-06
3.80	Trad	1.310e-03	1.310e-03	6.616e-19
	Cat	4.724e-19	4.724e-19	0.0
3.84	Trad	7.113e-04	7.113e-04	2.206e-19
	Cat	6.065e-19	6.065e-19	9.794e-35

Discussion

An interesting feature to note in the logistic experiment results is the presence of flat-line averages for the traditional grammar in each of the problem instances. What this demonstrates is a difficulty on the part of the traditional approach to evolve real numbers within its range or to alter expressed values in amounts of less than one. When the grammar produced a reasonably good fitness at the first generation, it settled upon this local minima. New expressions brought about through crossover and mutation failed to alter in small amounts the fitness of the best individual by bringing its value closer

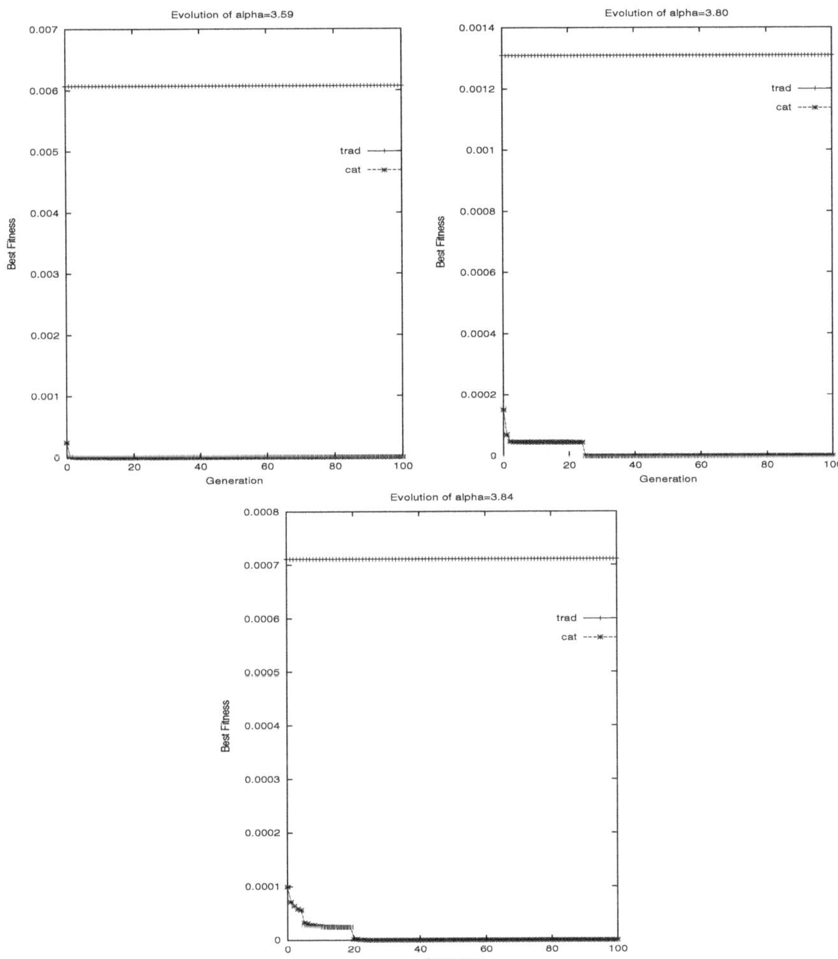

Fig. 5.3 Mean best fitness values (lower values are better) plotted against generations for each of the three grammars. Target values of α are 3.59 (top left), 3.80 (top right), and 3.84 (bottom).

to that of the target through different expressions. In order to simplify analysis, take, for example, the average fitness attained for the static experiment with target 5.67 (see Table 5.1). The average best fitness for this problem with the Traditional approach to constant creation was 0.33, with a standard deviation of 0 resulting in a flat-lined fitness series, meaning that each run settled upon a solution of 6 as the best phenotype. The problem encountered by the traditional approach here is that using the digits zero to nine and the operators provided, the task of adapting a relatively fit individual with one codon and a phenotype of 6 to 5.67 is too complex. In order to arrive at the correct phenotypic solution, an individual involving a number of extra codons would have to be evolved to produce an expression that resulted in the correct value. Slight deviations from a correct expression, or indeed the phenotype of 6, would produce fitnesses inferior to the simple one-codon solution and stand a lesser chance of being selected for reproduction. On the other hand, the concatenation approach can build upon a one-codon solution by concatenating the decimal point and other digits to arrive at the correct solution. As the results suggest, this was attainable on each run of the system. Digit Concatenation presents an incremental approach for the evolutionary search where the fitness may improve with the addition of each extra codon demonstrating a greater level of evolvability. Considering this, it can be said that while it is good at attaining a reasonable fitness for real targets within its range, forming expressions using integers to get real values proves too complex for the Traditional approach as it is less evolvable.

The Digit Concatenation approach, however, proved to be successful in evolving the correct solution for the static experiments across all runs. In the dynamic Markov type changes and the logistic difference equation type change good results were also arrived at.

Reflecting on the Traditional approach, these results would also suggest that the provision of a larger set of real values in the grammar might enhance the performance of this approach in its own right. The ability to mix different constant types illustrates the flexibility of the grammar-based approach to GP. To this end, a version of Ephemeral Random Constants for GE is investigated in the following section.

5.4 Analysis of Digit Concatenation and Persistent Random Constants

In Section 5.3, the Digit Concatenation method displayed superior performance over the Traditional technique for evolving constants in GE. In this section the Digit Concatenation method is analysed further by examining the preferences of evolutionary search when a number of different grammar-based constant generation methods are provided to GE. Along with Digit Concatenation, a novel grammar, defined as the Persistent Random Constants technique, is explored as well as the Traditional technique described in the previous section. All three methods are included in a grammar that only allows the use of one method exclusively. The preference of the evolutionary search is then examined across a range of constant generation problems.

5.4.1 Persistent Random Constant Creation in GE

In Section 5.2, a description of Ephemeral Random Constants in GP was given. Here, a form of ERC is introduced known as Persistent Random

Constants (PRC). Like the GP approach to Ephemeral Random Constants, the grammatical approach also generates a number of real values within a pre-specified range. Where it differs is that these numbers are then added to the grammar to be used by GE. This has the added effect that the random numbers become available to the evolutionary process throughout the lifetime of the experiment as they are part of the grammar itself. In GP Ephemeral Random Constants, these numbers, once evolved out of the population, cannot be re-introduced into the population. This can lead to a potential loss in the diversity of numbers available. Indeed, this approach does have similarities to the Traditional method for GE, where, instead of a small, fixed set of constants, PRC uses a larger number of randomly generated constants that persist for the lifetime of the run with the implication of better coverage of the constant search space.

5.4.2 Experimental Approach

A comparison is performed on the utility of three different constant creation methods for evolving constants by performance analysis on three different types of constant creation problems. The problems tackled are: finding a static integer, finding dynamic real constants, and finding a coefficient for the logistic difference equation. The problems used in this section, though similar to those in Section 5.3, use different targets. The reason behind this is to get a spread of targets both inside and outside the range of the Persistent Random Constants range. Most of the targets in Section 5.3 resided within this range. This is done to examine further the issue discussed in Section 5.3.4, where the Traditional approach was observed to settle on local minima, and also to examine the Persistent Random Constant's ability to evolve targets outside its range. These three methods of constant creation are combined in one competitive combination grammar which selects one method exclusively for an individual. In conjunction with experiments using this grammar, experiments are also conducted that use each method on its own in a grammar, allowing comparative benchmarks to be drawn from the previous section.

Finding a Static Constant

The aim of this problem, as before, is to evolve a single integer constant in order to verify the functionality of the proposed grammar. For these experiments, two constants were selected: a simple integer value within the range of the Persistent random constants, 50; and a complex floating-point real number outside the range of the Persistent Random Constants, 20021.11501. Fitness in these experiments is the absolute difference between the target and evolved values, the goal being to minimise the difference value.

Finding Dynamic Real Constants

This instance of finding dynamic real constants involves a dynamic fitness function that changes its target real constant values at regular intervals (every 10th generation). Two instances of this problem are tackled: the first sets the successive target values to be 192.47, 71.84, 173.59, and 192.47; the second instance oscillates between the two values 192.47 and 71.84, presenting Markov and Deterministic type problems as in the previous section. The aim here, as in the previous section, is to analyse the different constant representations in terms of their ability to adapt to a changing environment, and to investigate that behaviour in the event of both small and large changes. As in the static constant problem, fitness in this case is the absolute difference between the target and the evolved values, with the goal being the minimisation of this difference value.

The Logistic Difference Equation

This problem is used in the same manner with the same parameters as in Section 5.3.

Constant Creation Grammar

Three constant generation techniques are employed within the same grammar provided below.

```
<exp> ::= <value>
<value> ::= <trad> | <catR> | <persistent>
<op> ::= + | - | / | *
<trad> ::= <trad> <op> <trad> | <tradT>
<tradT> ::= 0 | 1 | 2 | 3 | 4 | 5 | 6 | 7 | 8 | 9
<catR> ::= <cat> <dot> <cat> | <cat>
<cat> ::= <cat> <catT> | <catT>
<catT> ::= 0 | 1 | 2 | 3 | 4 | 5 | 6 | 7 | 8 | 9
<dot> ::= .
<persistent> ::= <persistent> <op> <persistent> | <persistentT>
<persistentT> ::= ``150 randomly generated real constants''
```

The concatenation part of the grammar (`<catR>`) only allows the creation of constants through the concatenation of digits and the dot character for real numbers. This is in contrast to the Traditional part of the grammar (`<trad>`) that restricts constant creation to the generation of values from expressions using a fixed set of constants specified by the non-terminal `<tradT>`. The third part of the grammar concerns Persistent Random Constants. In this method, a set of 150 real-valued constants are generated randomly in the range 0 to 100 (inclusive) at the outset of a run. These are then directly

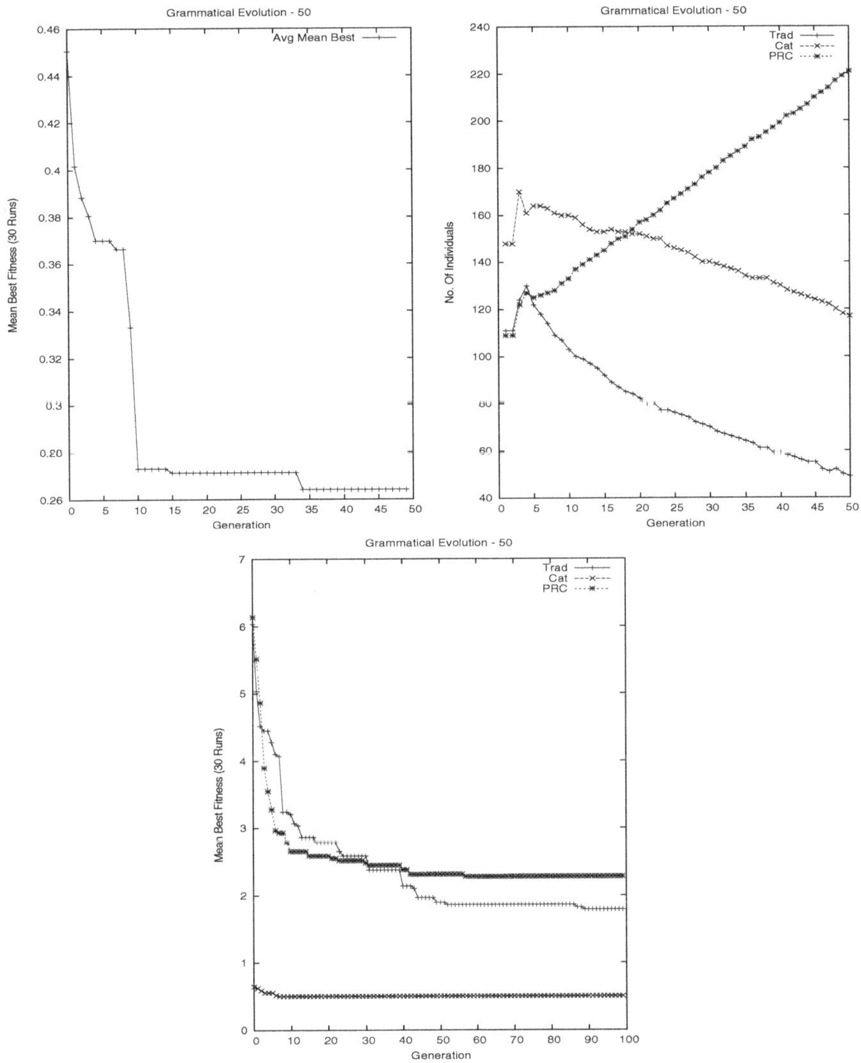

Fig. 5.4 Mean best fitness values (lower values are better) plotted against generations (top left), the number of individuals that use each of the three constant generation methods (top right), and a comparison of the performance of the exclusive component grammars (bottom)

incorporated as choices for the non-terminal <persistentT>. In a standard GP manner, these constants can then be utilised in arithmetic expressions to generate new constant values. The <value> production then is essentially the rule that permits the exclusive choice of one of these methods for each individual.

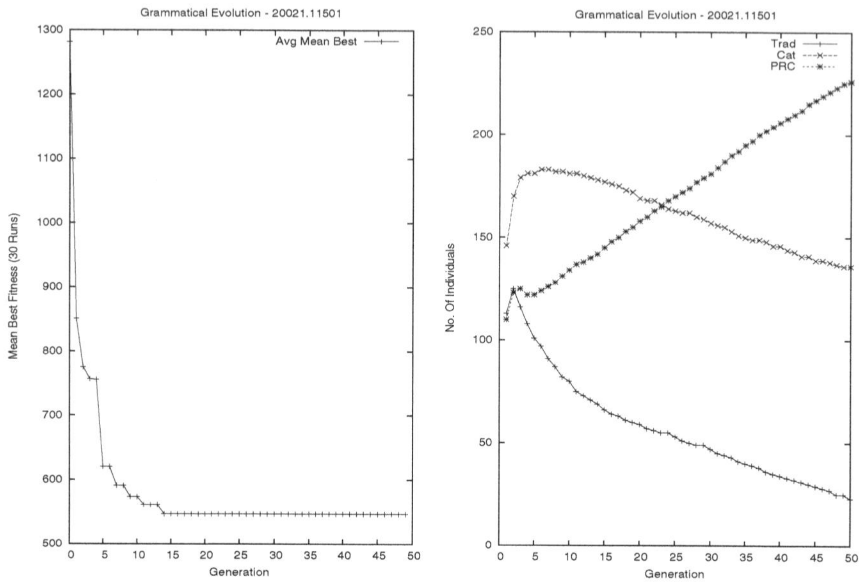

Fig. 5.5 Mean best fitness values (lower values are better) plotted against generations (left) and the number of individuals that use each of the three constant generation methods (right)

5.4.3 Results

For every problem instance, the parameters used and the number of runs conducted were the same as in Section 5.3.4.

Finding a Static Constant

For evolving the static constant 50, the results presented in Figure 5.4 indicate a preference by GE for the PRC in this problem. By the final generation, on average across thirty runs, GE evolved 221 individuals using the PRC method, against 117 and 59 for the Digit Concatenation and the Traditional methods respectively. Out of the 30 runs conducted of the best-performing individuals in the final generation, 60% had evolved a Concatenation individual and 40% a PRC individual.

Among the experiments incorporating each constant creation method exclusively as presented in Figure 5.4, the benefits of the Digit Concatenation method are highlighted for this problem. Over the course of 30, runs Digit Concatenation produced best performers with an average fitness of 0.50024 compared against 1.7931 and 2.27586 for the Traditional and PRC methods respectively.

Figure 5.5 presents the results for evolving 20021.11501. Here PRC is again seen to grow and dominate the population with 226 members against 136

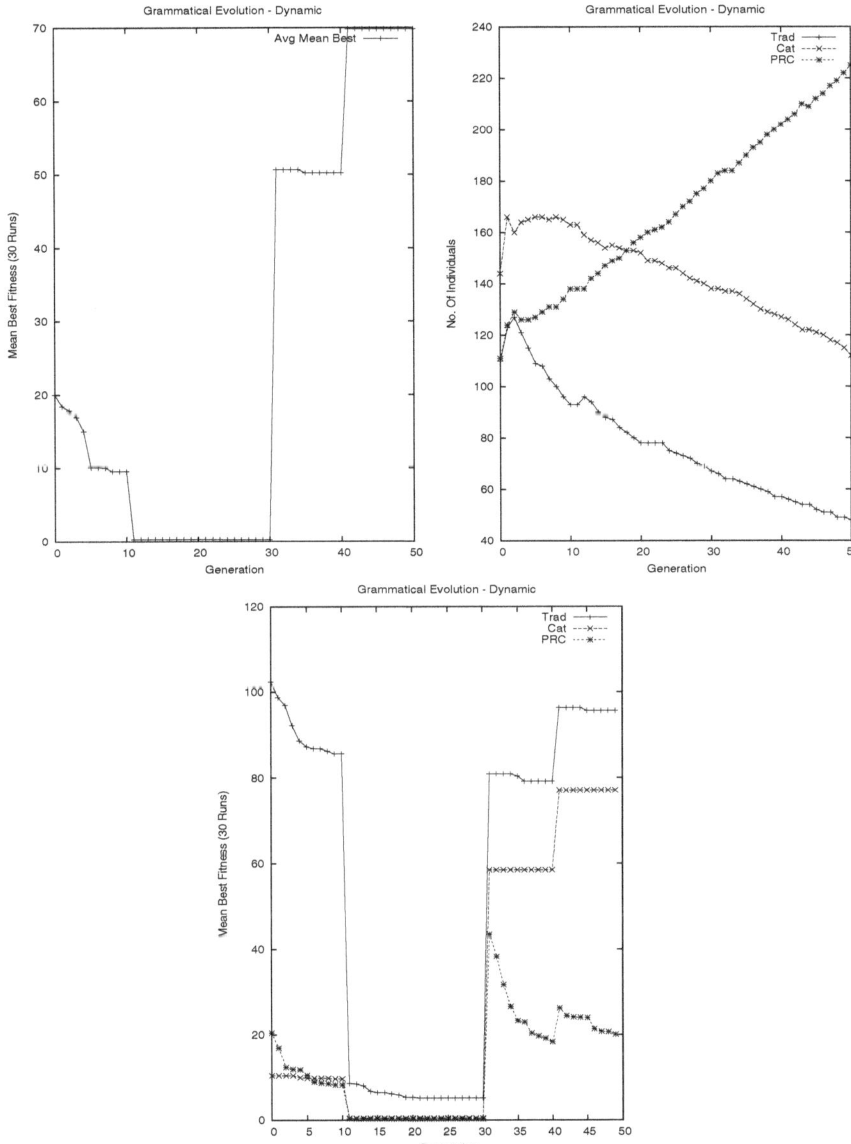

Fig. 5.6 Mean best fitness values (lower values are better) plotted against generations (top left), the number of individuals that use each of the three constant generation methods (top right), and a comparison of the performance of the exclusive component grammars (bottom)

and 23 for the Digit Concatenation and Traditional methods, respectively. However, in this instance Digit Concatenation is the method used for 100% of the best individuals, yielding an average best performance of 547.217.

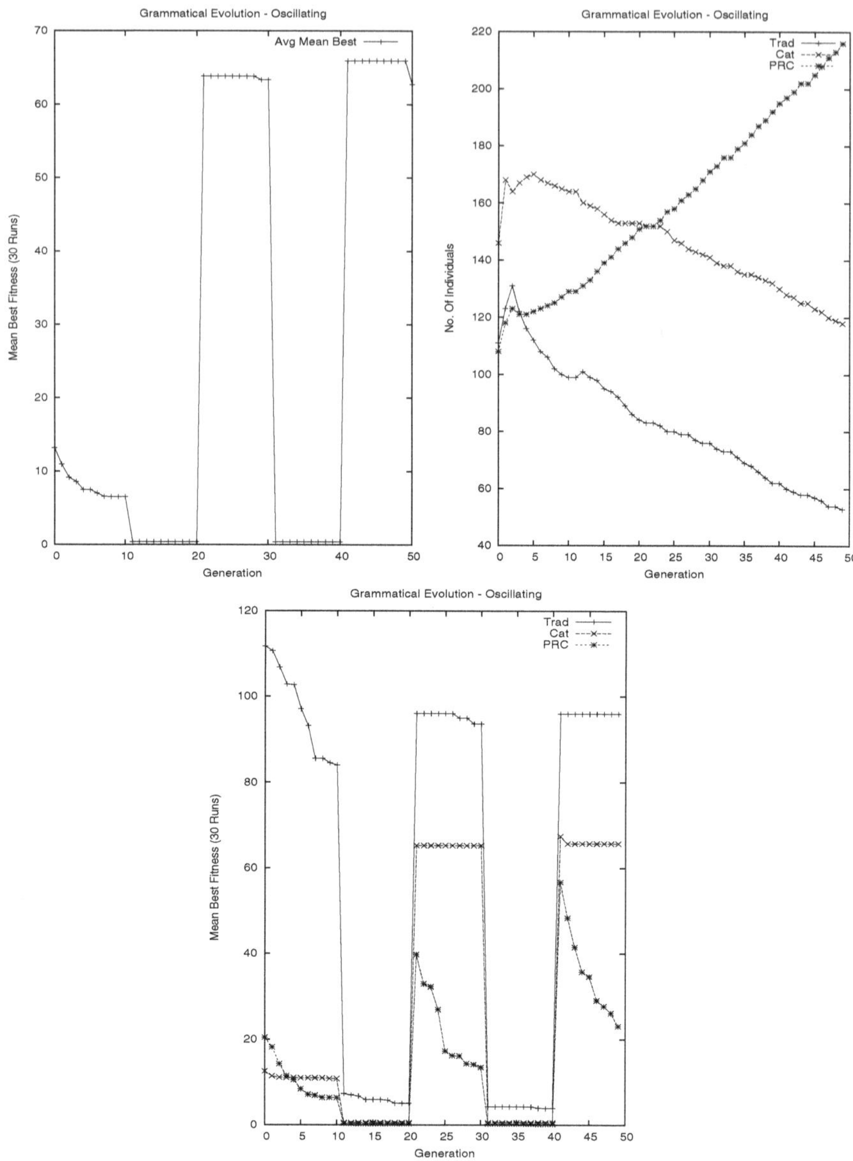

Fig. 5.7 Mean best fitness values (lower values are better) plotted against generations (top left), the number of individuals that use each of the three constant generation methods (top right), and a comparison of the performance of the exclusive component grammars (bottom)

In the experiments with exclusive grammars, the Digit Concatenation method was seen to provide the best average fitness at 1005.24, with the PRC method providing an average best fitness of 10070.5.

Finding Dynamic Real Constants

In Figure 5.6, graphs are presented for the experiments where the set of numbers to be evolved over the course of a run are: 192.47, 71.84, 71.83, 173.59 and 192.47. This time PRC gains a stronger foothold in the population over the course of the run, overtaking Digit Concatenation before generation 20 and at the same time presenting good fitness. However, at generation 30, where the target changes to 173.59, this fitness deteriorates significantly and remains poor. This suggests that while the target was within the range of the PRC, it was able to quickly attain a high fitness and a strong position in the population. However, it was unable to successfully evolve from this position once the target left its range.

In the single method grammars, however, the PRC method does express a stronger ability to evolve to the targets outside its range, taking large evolutionary steps towards the target after its initial change. The Digit Concatenation and Traditional methods present performances similar to the combination grammar's performance.

Results for the oscillating non-stationary problem instance are presented in Figure 5.7. In the second instance of this problem, where the target oscillates from 192.47 to 71.84 every 10 generations, a similar trend is noticed. Again by generation 20, PRC has reached a strong position within the population after a period with 71.84 as the target. The fitness drops drastically when the target changes to 192.47. When the target reaches the higher number for the third time, the fitness is worse again due perhaps to a further loss of diversity in the population.

As with the single grammars in the dynamic problem, the results for the oscillation experiments provide similar performances, with the PRC method being able to take the larger evolutionary steps once the target changes.

The Logistic Difference Equation

Figure 5.8, presents a sample of the results for the logistic difference equation with α values of 3.59, 3.8, and 3.84, which were very similar across each of the values. Here the Digit Concatenation method gains the dominant position within the population as evolution progresses. The proportion of PRC individuals in the population is seen to approach the level of Digit Concatenation constants initially, as in the dynamic experiments. However, this time, as evolution progresses, the Digit Concatenation method gains the dominant position within the population. Among the best performing individuals for 3.59, 60% were PRC; and 40% were Digit Concatenation; for 3.8, 73% were Digit Concatenation and 27% were PRC and for 3.84, 80% of best individuals were Digit Concatenation with 20% were PRC. No Traditional individuals were the best performer in any test.

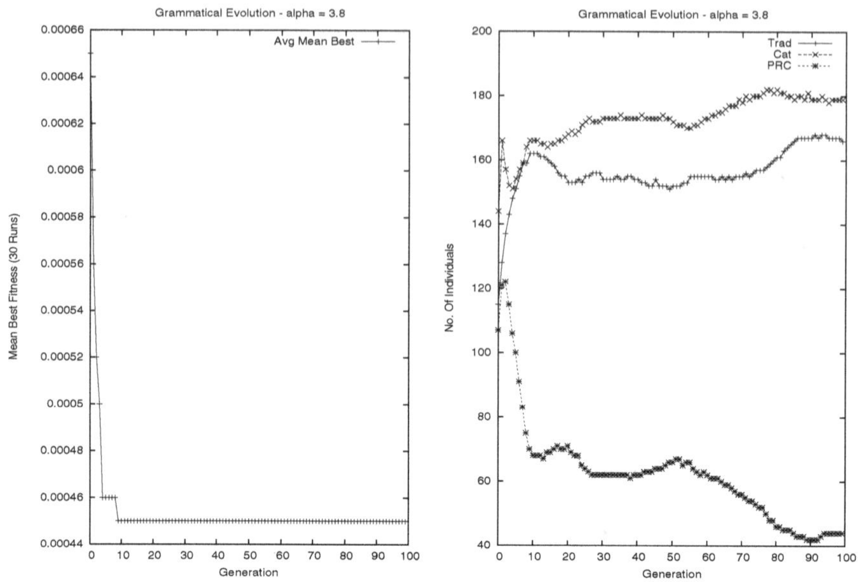

Fig. 5.8 Mean best fitness values (lower values are better) plotted against generations (left), and the number of individuals that use each of the three constant generation methods (right)

Discussion

Table 5.5 presents a summary of the the results from these experiments. A common feature seen in almost all the results across the problems examined in this section, with the exception of the logistic difference equation problem, shows the Persistent Random Constants method consistently gaining larger portions within the population, while not seeing this population dominance reflected in the number of best-performing solutions. For the problems of evolving the static integer 50 and the dynamic targets, numbers are presented that are within the range of the Persistent random constants. Considering this, the reason why the Persistent Random Constants method gains such a large portion is likely to be that this method would contain a relatively large number of terminals within the neighbourhood of the target, reflecting a similar advantage to the Traditional method in Section 5.3.4. Thus this method presents the evolutionary process with again a relatively large number of simple solutions, or terminals, with good fitnesses in proportion to the other two methods in the grammar. In the dynamic problems, once the target moves outside its range, the PRC method had already attained a strong postion in the population, gaining a certain reproductive momentum to the detriment of the Digit Concatenation and Traditional methods. In terms of evolvability the PRC method would also be able to attain reasonable fitnesses simply

Table 5.5 Summary of results including the percentage of population occupied by each method and the percentage of best individuals of each type by final generation. Note: The percentage of population figures do not add up to 100% due to individuals not mapping.

Exp	Metric	Trad	Cat	PRC
50	**Pop %**	12	23	**44**
	Best %	0	**60**	40
20021.11501	**Pop %**	5	27	**45**
	Best %	0	**100**	0
Dyn	**Pop %**	8	24	**43**
Osc	**Pop %**	7	23	**43**
3.59	**Pop %**	31	**37**	10
	Best %	0	40	**60**
3.8	**Pop %**	33	**35**	9
	Best %	0	**73**	27
3.84	**Pop %**	33	**36**	9
	Best %	0	**80**	20

by summing up two of its number terminals at the higher end of its range, requiring the exploitation of just three terminals or codons.

This explanation can then be related directly to the good performance seen by the Traditional method in attaining a larger proportion of the population in the logistic difference equation problems and entirely contrasting its performance in the other problems. In the logistic difference equation problem the target was within the range of the Traditional grammar and also within the range of PRC. The difference, however, is that while it was within both their ranges, the Traditional method presented the evolutionary process with a much smaller selection of terminals (the 10 digits), all of which would have had reasonable fitness on their own, compared to the PRC method, which had 150 terminals dispersed evenly from 0 to 100. The Traditional approach presented a more evolvable set of choices for this specific problem. What held the Traditional method back in this case, however, is its difficulty in evolving fitter solutions that require small changes in the resultant value of a best-performer's expression. This problem was already discussed in Section 5.3.4. The Digit Concatenation method, on the other hand, offered greater evolvability in attaining fitter solutions in Section 5.3.4 for this problem, explaining how it could hold onto a majority position within the population on average.

In examining the static problem with target 20021.11501, a different situation is presented. In this problem the target was well outside the range of the PRC method yet it still grew to dominate the population. The behaviour of Digit Concatenation trend line in this instance, however, differs from the other problems. Here Digit Concatenation ends up with a proportion of the population that is just 12 less than what it began with on average. Considering that it provided 100% of the best-performing solutions, evolutionary pressure allowed it to maintain its population share. In the three-way evolutionary competition to gain population share the PRC method then grew its population share, to the detriment of the Traditional method. The Traditional method struggled to cope with the ease at which the PRC method could evolve much larger values with expressions involving its larger terminal values. Between the Traditional approach and PRC, the PRC method is seen to be more evolvable because it can sum or multiply out these larger terminals. PRC also has this advantage over the Digit Concatenation approach. However, because Digit Concatenation could accurately evolve the target, selection pressure allowed it to maintain its population share.

In summary, it can be seen that PRC presents an ability to take large evolutionary steps towards a target, while Digit Concatenation proved to be the most accurate in evolving its targets. The way in which both methods evolve their targets gives insight to this behaviour. PRC was able to achieve a high proportion of the population by evolving reasonably fit solutions with a lower number of codons through the use of expressions. Digit Concatenation, in comparison, had to construct a solution with the correct number of digits with a left-to-right dependency, whereby the first digit had the greatest impact on fitness. Having discovered a fit neighbourhood, Digit Concatenation then allowed a more finely grained approach to evolving the correct target accurately.

Both Digit Concatenation and PRC present relative advantages and disadvantages over each other: Digit Concatenation is accurate in evolving both large and small numbers though more slowly evolving while PRC is fast in attaining a good fitness though not as accurate. These are advantages that are desirable in constant generations mechanisms where dynamic environments are concerned. PRC has demonstrated an ability to quickly discover a fit neighbourhood, while Digit Concatenation can fine-tune a solution. Both advantages can be examined in terms of evolvability. While PRC can more easily evolve to a fit neighbourhood through multiplying or summing large terminals, Digit Concatenation can conduct evolutionary perturbation to more accurately find the target. However Digit Concatenation suffers from another weakness in that it cannot form a phenotype to produce the same value as certain fractions, for example a 1/3, as it cannot concatenate digits to infinity. The next section will seek to address the evolvability weakness in the Digit Concatenation approach along with its inability to form factions by equipping it with an ability to form expressions and also exploring a grammar that allows both methods to cooperate.

5.5 Direct Comparison of Digit Concatenation and Persistent Random Constants

Section 5.4 demonstrates the superiority of both the Digit Concatenation and Persistent Random Constant methods over the Traditional approach. In order to gain a more accurate understanding of the relative advantages of these two methods, and the merits of a combination of both approaches, a further series of experiments were undertaken. This section compares the two methods using grammars similar to the previous section, along with grammars which use each approach exclusively. However, in these experiments the Digit Concatenation method is additionally given the ability to form expressions so that it can mimic the ability of the PRC approach to more easily evolve individuals to fit neighbourhoods. The mathematical operators adopted are again consistent with earlier experiments.

Below are the combination grammars derived from experiments in the previous section. The first is most similar to the previous section, except that it only uses the Digit Concatenation method and Persistent Random Constants. The second grammar presents GE with a method to ascertain whether the two approaches may mutually complement each other, as this (cooperative) grammar allows the formation of expressions using constants derived from both paradigms.

Competitive Grammar

```
<exp> ::= <number>
<number> ::= <catE> | <persistent>
<op> ::= + | - | / | *
<catE> ::= <catE> <op> <catE> | <catR>
<catR> ::= <cat> <dot> <cat> | <cat>
<cat> ::= <cat> <catT> | <catT>
<catT> ::= 0 | 1 | 2 | 3 | 4 | 5 | 6 | 7 | 8 | 9
<dot> ::= .
<persistent> ::= <persistent> <op> <persistent> | <persistentT>
<persistentT> ::= ''150 randomly generated real constants''
```

Cooperative Grammar

```
<exp> ::= <number>
<number> ::= <value> <op> <value> | <value>
<value> ::= <catE> | <persistent>
<op> ::= + | - | / | *
<catE> ::= <catE> <op> <catE> | <catR>
<catR> ::= <cat> <dot> <cat> | <cat>
<cat> ::= <cat> <catT> | <catT>
<catT> ::= 0 | 1 | 2 | 3 | 4 | 5 | 6 | 7 | 8 | 9
<dot> ::= .
<persistent> ::= <persistent> <op> <persistent> | <persistentT>
<persistentT> ::= ''150 randomly generated real constants''
```

The following grammars incorporate each method for constant creation exclusively and are designed for a symbolic regression problem to evolve the equation for the area of a circle.

Exclusive Cat

```
<exp> ::= <catE>
<op> ::= + | - | / | *
<catE> ::= <catE> <op> <catE> | (<catE><op><catE>) | <catR>
<catR> ::= <cat> <dot> <cat> | <cat>
<cat> ::= <cat> <catT> | <catT>
<catT> ::= 0 | 1 | 2 | 3 | 4 | 5 | 6 | 7 | 8 | 9
<dot> ::= .
```

Exclusive Persistent

```
<exp> ::= <prc>
<op> ::= + | - | / | *
<prc> ::= <prc> <op> <prc> | (<prc><op><prc>) | <prcT>
<prcT> ::= ``150 randomly generated real constants''
```

5.5.1 Experimental Approach

The experiments performed focus on three areas: one being the creation of a large complex number outside the range of the Persistent Random Constants; on the flexibility of the methods in a dynamic environment; and on evolving the form and constant, π, in the equation for calculating the area of a circle. This final experiment differs from previous experiments as having examined GE's ability to create and adapt constants, its ability to evolve an equation's form is now also examined.

Finding a Static Constant

The target of 20021.11501 was again chosen for these experiments to enable direct comparisons with previous sections and also because of its difficulty, as it represents a high precision floating-point number outside the range of the Persistent Random Constants.

Finding Dynamic Real Constants

The same series of dynamic real constants, as used in section 5.4, are again considered; successive targets of 192.47, 71.84, 173.59, and 192.47 changing at the 10th generation for fifty generations. Once more, the ability to compare results with the previous section is provided.

Finding the Equation for the Area of a Circle

For this section, a new problem is introduced. In this case two grammars are used that incorporate each method exclusively, with the aim of evolving the equation of a circle, πr^2. The setups are tested against 100 radii each generation with the range $2 \rightarrow 102$, where the objective is to minimise the cumulative difference to the correct area across the 100 radii.

5.5.2 Results

For every problem instance, the parameters used and number of runs conducted were the same as in Section 5.3.4.

Finding a Static Constant

In this case, Figure 5.9 demonstrates that the Digit Concatenation method began to gain an upper hand on average within the populations at generation 13. Finishing at the final generation with a large majority of the population, Digit Concatenation has 318 versus 94 for the Persistent Random Constants method. Of the best-performers, only 1 of the 30 runs provided a solution using the Persistent Random Constants method with the best Digit

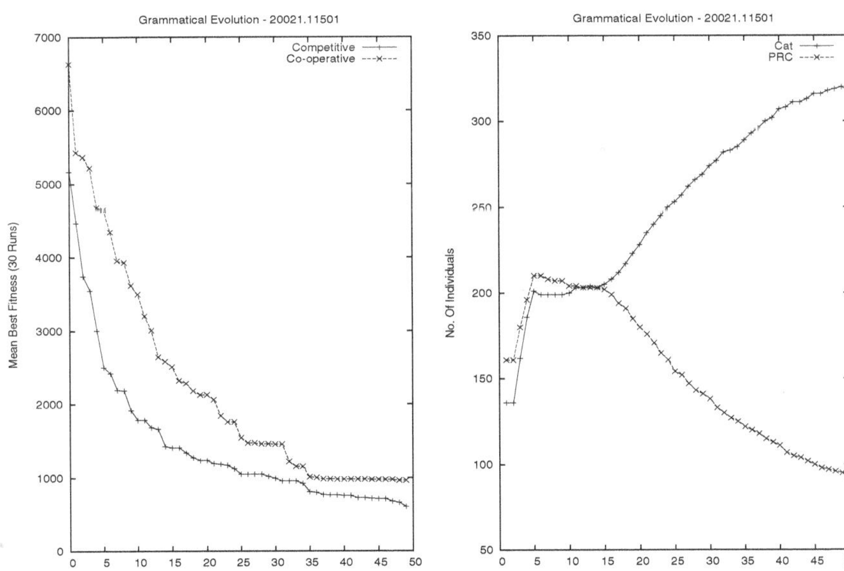

Fig. 5.9 Mean best fitness values (lower values are better) plotted against generations (left), the number of individuals that use each the constant generation methods (right) for the competitive grammar

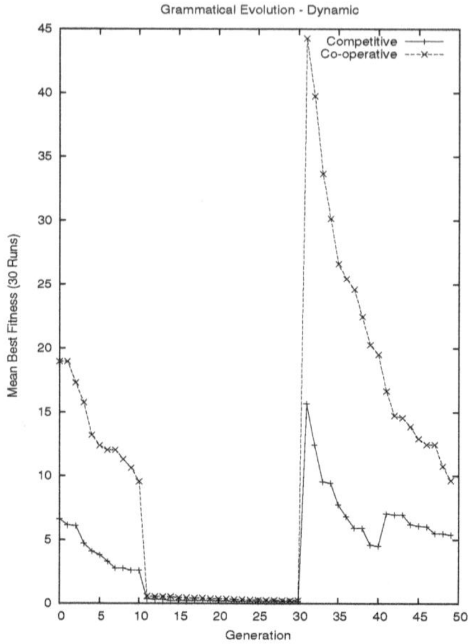

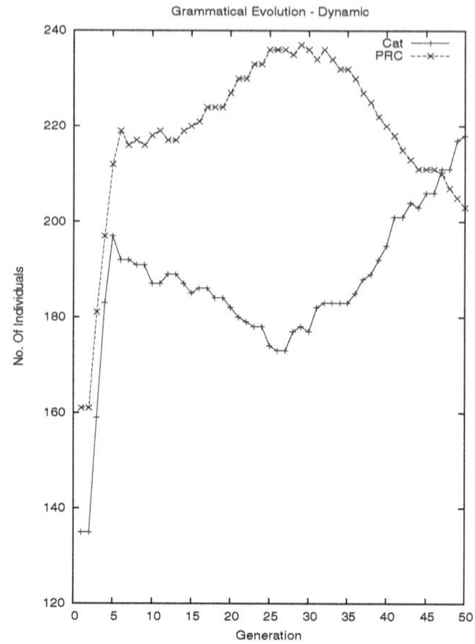

Fig. 5.10 Mean best fitness values (lower values are better) plotted against generations (left) and the number of individuals that use each of the constant generation methods (right)

Concatenation solution producing an expression that came to within 18.3872 of the solution. This solution is provided below.

```
20002 + 0.727829
```

Interestingly, the decimal and whole part are separated out into an expression. The whole part is also evolved on the more significant side of the phenotype according to GE's mapping process with the decimal part summed in on the right. By the final generation, the best performer on average produced a fitness of 607.968. Among the experiments with the complementary grammar, the average best fitness by the final generation was a comparable 688.798, with a t-test and bootstrap t-test demonstrating no statistical difference between the results.

Finding Dynamic Real Constants

Figure 5.10 displays the results for the dynamic experiments. A similar trend to that seen in Section 5.4.3 is presented. Again, the Persistent Random Constants method gains a stronger position within the population while the target is within its range. The difference here is that once the target leaves this range, the the Digit Concatenation method begins to gain a bigger share of the population and ends up with a slight majority at 218 to 203. It can also be noted that a higher rate of evolution occurs in these experiments

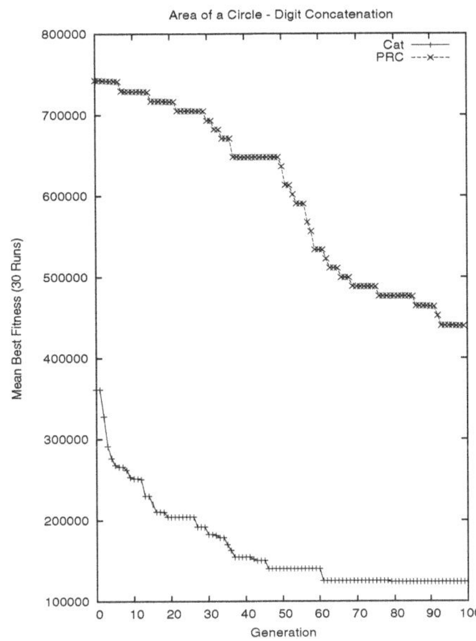

Fig. 5.11 Mean best fitness values (lower values are better) plotted against generations, where fitness is the cumulative difference of each individual for 100 radii to the correct area

when the target goes outside the Persistent Random Constants range. This combined with the higher frequency of Digit Concatenation individuals within the population would suggest that the ability for the Digit Concatenation method to create expressions is directly responsible for the improvement in the rate of evolution across both grammars.

Comparative analysis of the grammars at generations 10, 20, 30, 40 and 50 using a t-test and bootstrap t-test reveal a statistical significance in the difference in results at generations 10 and 40. No other transition generations showed a statistically significant difference.

Finding the Equation for the Area of a Circle

In Figure 5.11, it can be seen that the Digit Concatenation produces superior fitness over Persistent Random Constants. By the final generation, Digit Concatenation produces an average fitness of 12489 compared to 439226 for Persistent Random Constants.

Discussion

These experiments have focused on the direct comparison of the Digit Concatenation and Persistent Random Constants methods, while introducing an extra feature to Digit Concatenation; the ability to create and evolve expressions in conjunction with the numbers. This added feature is beneficial

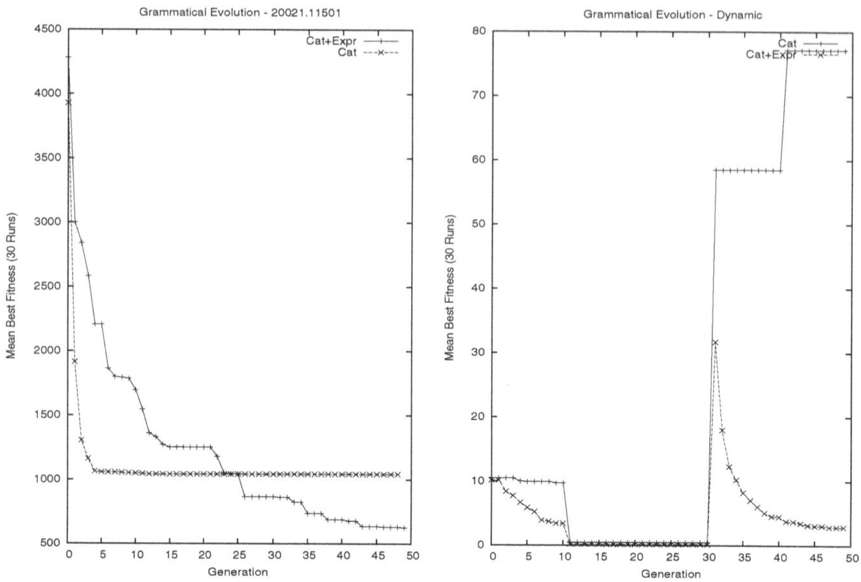

Fig. 5.12 Mean best fitness values (lower values are better) plotted against generations for 20021.11501 (left) and the dynamic experiment (right)

to Digit Concatenation. As this is the essential difference between the experiments both here and in Section 5.4, the results suggest that allowing Digit Concatenation to produce expressions enables it to achieve a higher degree of accuracy over the prior implementation. It also allows it to be more flexible in a dynamic environment, with the results for the dynamic experiment clearly outperforming those in Section 5.4 with positive evolutionary steps being taken at each generation after a target change. In order to gain direct perspective of the improvement in Digit Concatenation when it is provided with the ability to create expressions Figure 5.12 provides graphs that compare Digit Concatenation using expressions with pure Digit Concatenation. In these graphs, an improvement in performance over the pure Digit Concatenation results is seen in both the static problem and the dynamic problem.

5.6 Digit Representation and Problem Difficulty

In a recent series of experiments we turned to a tunably difficult problem, Binomial-3, as analysed for standard tree-based Genetic Programming by Daida et al [48]. Binomial-3 can be cast as a symbolic regression problem instance where the target function could take the following form:

$$1 + 3x + 3x^2 + x^3$$

There are a number of alternative representations for this target including:

$$(x+1)/(1/(1+(x/5)+(x/(1/x))))$$

$$(1+x)(1+2x+x^2)$$

$$1+x+x+x+x^2+x^2+x^2+x^3$$

Daida et al have demonstrated that this problem becomes exponentially more difficult for GP with Ephemeral Random Constants as the range of constants made available to GP increases. We wish to analyse the performance of the Digit Concatenation and Persistent Random Constant (PRC) representations on this problem to determine if they confer some advantage. The general form of the grammar adopted for this problem is presented below.

```
<expr> ::= (<op><expr><expr>) | <var> | <const>
<op> ::= + | - | *
<var> ::= x
<const>::= "constant generation method"
```

In the case of PRC 100 randomly generated constant values are generated in the allowed range. For Digit Concatenation, the grammars were range-specific, and an example of the grammars adopted is provided here.

```
For the range [0,5]
<int> ::= <nzdigit>.<digit><digit>
<digit> ::= 1|2|3|4|5|6|7|8|9|0
<nzdigit> ::= 0|1|2|3|4

For the range [0,100]
<int> ::= <nzdigit><digit>.<digit><digit>
        | <digit>.<digit><digit>
<digit> ::= 1|2|3|4|5|6|7|8|9|0
<nzdigit> ::= 1|2|3|4|5|6|7|8|9

For the range [0,5000]
<int> ::=<anzdigit><digit><digit><digit>.<digit><digit>
        | <nzdigit><digit><digit>.<digit><digit>
        | <nzdigit><digit>.<digit><digit>
        | <digit>.<digit><digit>
<digit> ::= 1|2|3|4|5|6|7|8|9|0
<nzdigit> ::= 1|2|3|4|5|6|7|8|9
<anzdigit> ::= 1|2|3|4
```

The experiments were implemented using GEVA [76] with the evolutionary parameter settings of population size 500, crossover rate 0.9, replication rate 0.1, maximum generations 200, roulette wheel selection and generational replacement, ramped-half-and-half initialisation with max derivation tree depth

Table 5.6 Results for Persistent Random Constants on the Binomial-3 problem

Case	BestFitness	stdDev	Average Fitness	stdDev
PRC5	9.56	10.9	4596.1	15122.4
PRC10	16.6	20.4	2579.5	6858.4
PRC50	90.09	64.3	13976.9	43121.6
PRC100	104.7442	71.9	2624609	1437136
PRC500	140.6	57.7	1512553876	8276308040
PRC1000	164.3	65.04	2996614	13835151
PRC2000	164.1	50.8	586146.7	305680.9
PRC5000	160.9	62.2	8.716e+16	4.77e+17

Table 5.7 Results for Digit Concatenation on the Binomial-3 problem

Case	BestFitness	stdDev	Average Fitness	stdDev
Concat5	1.14	1.65	727.3	2709.2
Concat10	3.19	5.16	10654.27	34445.8
Concat50	4.5	6.88	52486.8	124798.6
Concat100	5.2	6.97	1418010	6980128
Concat500	6.15	8.22	84063967	409599805
Concat1000	8.2	8.68	901673203	4857220418
Concat2000	5.44	7.25	2815668301	10691830451
Concat5000	3.43	4.75	240268558871	1.311852e+12

limit 17. Thirty independent runs were performed for each constant representation on each range of constants values, which were [0,5], [0,10], [0,50], [0,100],[0,500], [0,1000], [0,2000] and [0,5000].

The results are striking in Digit Concatenation's ability to seemingly ignore the increasing problem difficulty as the range increases. A summary of the results is outlined in Tables 5.7 and 5.6 and Figures 5.13 and 5.14. As evidenced by these results on problem difficulty and the experimental evidence detailed earlier in this chapter, representation can have a profound impact on the performance of Genetic Programming. The cumulative evidence of this chapter tends to support the adoption of Digit Concatenation with expressions as the method of choice for constant creation in GE.

5.7 Conclusions

The ability to evolve new, and adapt existing, constants plays an important role in the effective use of GP paradigms for dynamic environments. When the environment undergoes change, the population of solutions needs to be able to draw upon its diversity, have the ability to evolve to the new circumstances, and explore new areas of the solution space. The objective of this chapter was to identify the best method for constant creation and adaptation

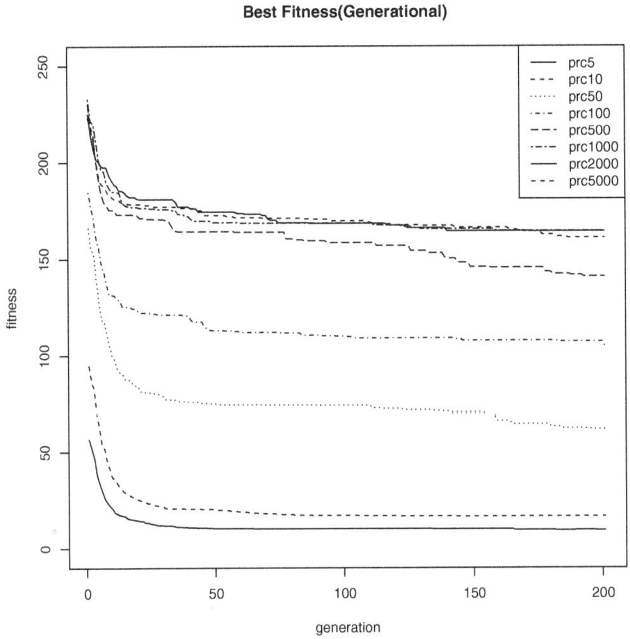

Fig. 5.13 PRC average best fitness over thirty runs on Binomial-3 with increasing range of constants

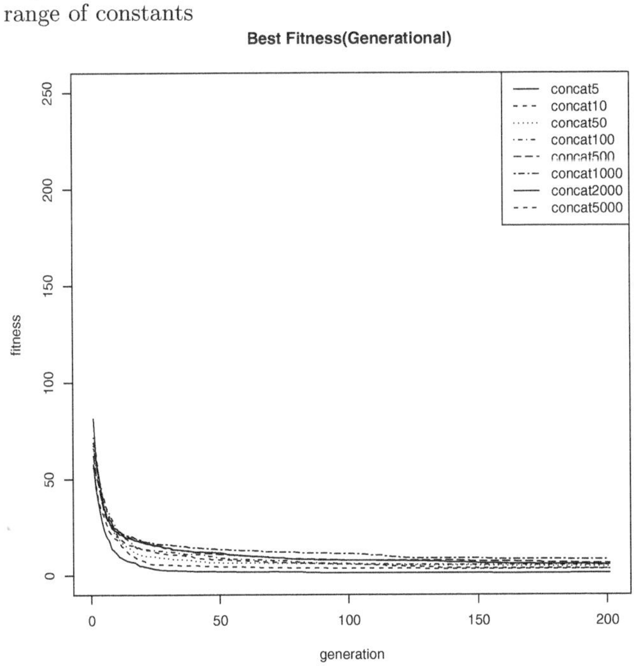

Fig. 5.14 Digit Concatenation average best fitness over thirty runs on Binomial-3 with increasing range of constants

with a focus on performance in dynamic environments, addressing the limitations of standard GP's ERC. To this end three methods in particular were examined on a series of benchmark problems; Traditional, Digit Concatenation and Persistent Random Constants. Throughout these experiments, the Digit Concatenation method was seen to produce the best results with more regularity than all other methods investigated. In Section 5.3, a pure Digit Concatenation grammar exhibited better performance across all of the problems.

Section 5.4 presented a combination grammar where the choice of the different methods was left up to the evolutionary process. Here, Persistent Random Constants were seen to grow and occupy the majority of populations, on average, in most experiments. However, this majority did not translate to PRC producing the best individuals, as Digit Concatenation produced the majority of best fit individuals. Among experiments with the exclusive grammars, the Digit Concatenation method provided superior fitness for the static problems again, with Persistent Random Constants giving better fitnesses for the dynamic problems. PRC thus demonstrated a greater level of evolvability in that it was able to shift to new solutions.

Section 5.5 sought to address the weakness in evolvability of the Digit Concatenation method by providing it with the ability to evolve expressions. This proved to significantly improve fitnesses in the dynamic problem and in contrast to Section 5.4, saw the Digit Concatenation method grow to take up a majority position within the population for the static problem from an early stage. An interesting story is told by the population graph for the dynamic problem, where the Persistent Random Constants method consumes a majority of the population while the target was within its range, but the trend peaks and reverses at generation 30 when the target moves outside of its range. The result by the final generation gives a slight majority to the Digit Concatenation method. In terms of evolvability, when the target is within the PRC range, method selection is biased towards PRC as it able to more easily evolve the target given the presence of terminals around it. This evolutionary edge is lost when the target transitions outside the PRC range and Digit Concatenation can acquire a majority. The experiments conducted in this section indicate that the ability to create expressions among evolved constants provides Digit Concatenation with a mechanism to shift its search neighbourhood with greater ease and take steps to another area of the solution space.

These experiments have examined different problems involving static constants both complex and simple; dynamic problems with large and small variations in targets; oscillating targets; a co-efficient to a chaotic equation; and symbolic regression. Considering the results, a constant creation grammar that provides the Digit Concatenation method with the ability to create expressions is the most advantageous method explored. Its ability to constantly introduce new constants to the system, take consistent evolutionary steps towards targets and produce a higher proportion of best-performing individuals in comparative tests mark it apart from the other methods explored. At this point, it is useful to compare the performance of Digit Concatenation combined with an ability to form expressions over these experiments, with the desireable features outlined at the beginning of the chapter:

i. **Static Targets:** Of all the methods tested, Digit Concatenation proved to be the most accurate at evolving static constants. While PRC demonstrated an ability to perform well in Section 5.4, once Digit Concatenation was also able to form expresions it presented superior performance.
ii. **Types of Change:** In these experiments, the mechanisms were tested against Markov type changes in the dynamic problem and Deterministic in the oscillating. While Digit Concatenation presented superior performance against the Traditional approach in Section 5.3, it did not present as good a performance in Section 5.4, where the targets were over a wider range. Its results improved considerably however once combined with the ability to form expressions and outperformed PRC in Section 5.5.
iii. **Adaptable:** Through out these experiments, Digit Concatenation provided the most accurate solutions demonstrating an ability to fine-tune its constants. However, it suffered if the target changed by a large amount. Section 5.5 solved this problem by allowing the method to form expressions between its evolved constants, enabling it to take large evolutionary steps towards a new fit neighbourhood.
iv. **New Constants:** Of the three mechanisms tested Digit Concatenation is the only one that has the ability to create entirely new constants. Both the Traditional and PRC methods begin with a fixed range of constant terminals in the grammar that is specified a-priori. From here, all other values are only arrived at through expressions on these constants. As the experiments suggest this does not provide solutions as accurate as through evolving the individual constants themselves.
v. **Large and Small Constants:** In experiments evolving the co-efficient for the logistic difference equation, the mechanisms were required to evolve a small number accurately. Again Digit Concatenation presented the most accurate results. In evolving the large number, Digit Concatenation also provided the most accurate results. However, without the ability to form expressions Digit Concatenation did not attain reasonable fitnesses as easily as PRC.

While the focus of these experiments was to identify an efficient and effective means of constant creation and adaptation for GE in dynamic environments a number of these experiments also provided insights into the behaviour of GE itself in dynamic environments. As the problems were relatively simple, fundamental features were more easily identified. Emerging from this evolvability was identified as being key. Providing GE with an ability to take large evolutionary steps initially along with an ability to fine tune the phenotypes allows it to produce more accurate results faster.

Considering the results, a constant creation grammar that provides the Digit Concatenation method with the ability to perform expressions is the most beneficial, and efficient method explored and simultaneously addresses the weaknesses of ERC in standard GP. These results are a practical finding for the efficient and effective use of GE in dynamic environments. Thus, future experimentation in this book under the financial domain will use this grammar for generating constants.

Chapter 6
Constant Creation with meta-Grammars

In the previous chapter, the utility of different variations of the Digit Concatenation method for constant creation were examined. Digit Concatenation was also combined with and compared to, Persistent Random Constants. In these experiments, the Digit Concatenation method was seen to produce the most fit individuals more regularly than the other methods examined. In this chapter, the Digit Concatenation method is further explored by examining Digit Concatenation with the use of a meta-Grammar based approach using Grammatical Evolution by Grammatical Evolution ($(GE)^2$) [156]. A meta-Grammar is employed in a dual-chromosomal structure, where one chromosome describes the solution as usual and the second chromosome is the individual's own grammar that maps the solution chromosome. The meta-Grammar is used to map the grammar chromosome for each individual. Providing GE with the ability to evolve its own grammar, specific to each solution, presents an extra layer of adaptability.

This added feature may have benefits for evolution in dynamic environments in particular, as it allows GE to evolve the very vocabulary it uses to describe phenotypic solutions, potentially incorporating biases into the grammar learned through feedback from the environment. This extra layer is a feature that is unique to GE over the fixed function and terminal sets allowed in GP. The inclusion of meta-Grammars also matches with the adaptability requirement for constant generation techniques for dynamic environments, outlined at the beginning of the previous chapter. Added to this, the solution grammar itself has the potential to form a representational memory where the type of change in a dynamic problem is Deterministic oscillatory. The solution grammar can achieve this through biasing the grammar towards evolving the targets that are being switched between.

In this chapter, the performance of the meta-Grammar approach is compared with that of the grammars in the previous chapter. The aim of the chapter is to determine whether the incorporation of a meta-Grammar and $(GE)^2$ produces a boost in performance over the standard GE approach. The next section begins with a brief description of the Grammatical Evolution

I. Dempsey et al.: Foundations in Gramm. Evol. for Dyn. Environ., SCI 194, pp. 105–120.
springerlink.com

by Grammatical Evolution paradigm to refresh the reader's memory. This is followed by two sections of experiments that examine the utility of $(GE)^2$ and seek to further understand the behaviour of the Digit Concatenation approach to constant creation.

6.1 Grammatical Evolution by Grammatical Evolution

A description of the Grammatical Evolution by Grammatical Evolution (GE^2) approach is provided in Chapter 4. A brief summary of the approach is provided here as a reminder. In order to allow evolution of a grammar, for $(GE)^2$, another grammar (meta-Grammar) must be provided to specify the form a grammar can take. This is an example of the richness in the expression of grammars that make the GE approach so powerful. By allowing an EA to adapt its representation (in this case through evolution of a solution's grammar), it provides the population with a mechanism or an extra layer of adaptability to survive in dynamic environments in particular, and also to automatically incorporate biases into the search process.

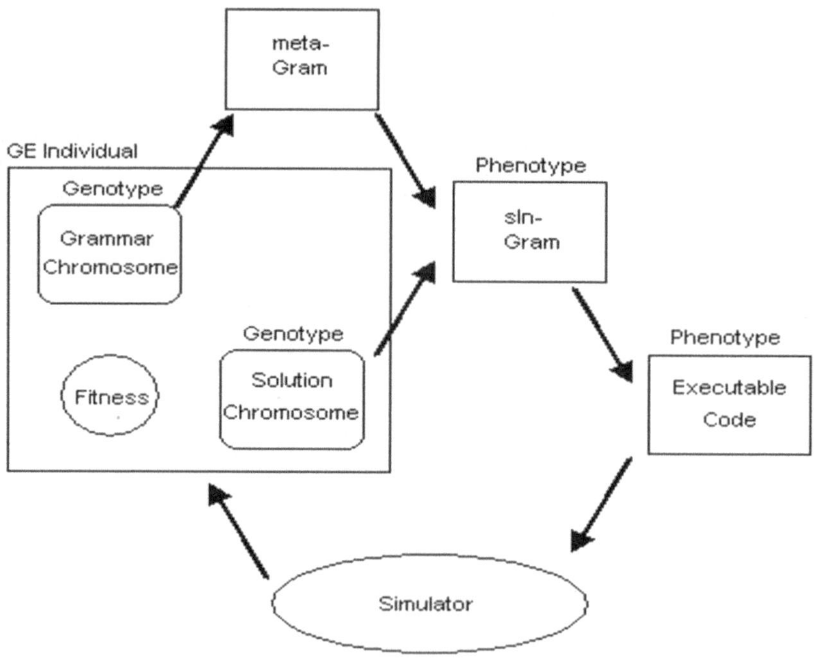

Fig. 6.1 The grammar genotype chromosome is mapped to its BNF phenotype using the meta-Grammar. The solution genotype is mapped to executable code using the solution grammar. This is evaluated in the simulator and a fitness returned to the GE individual that is made up of two genotypic chromosomes and a fitness score.

In this approach then, there are two distinct grammars: the *meta-Grammar* (or grammars' grammar) and the *solution grammar*. The meta-Grammar dictates the construction of the solution grammar. In the experiments described in this chapter, two separate variable-length, genotypic chromosomes were used: the first chromosome to generate the solution grammar from the meta-grammar, and the second chromosome to generate the solution from the solution grammar. Crossover operates between homologous chromosomes, that is, the solution-grammar chromosome from the first parent recombines with the solution-grammar chromosome from the second parent, with the same occurring for the meta-grammar chromosomes. In order for evolution to be successful it must co-evolve both the grammar and the structure of solutions based on the evolved genetic code. Figure 6.1 presents a diagram illustrating the structure of a $(GE)^2$ individual as well as the mapping process.

Other researchers have also indirectly examined the evolution of grammars with GE by evolving Lindenmayer-systems. Ortega [169] employed GE to evolve Lindenmyer grammars for the construction of fractal curves, while Hemberg and O'Reilly [87] also evolve Lindenmyer grammars for the generation of curved surfaces in Genr8. Outside of GE, Keller and Banzhaf [103] investigated the evolution of genetic code by allowing each individual to evolve its own mapping table for the genotype-to-phenotype mapping.

The inclusion and evolution of an individual's own solution grammar provides two potential benefits for conducting evolution in dynamic environments:

- **Adaptability:** By allowing an individual to evolve its own vocabulary an extra dimension of adaptability is provided. Once the phenotypic grammar has changed as a result of genetic operators, whole new areas of the phenotypic solution space can be opened up due to the provision of new terminals in the solution grammar.
- **Representational Memory:** The fitness of an individual in $(GE)^2$ is dependent upon an effective grammar being evolved that allows the solution chromosome to produce a fit phenotype. Where the problem domain is dynamic, this phenotype will have to change over time. For an individual to survive and maintain a good fitness, it is therefore incumbent upon the solution-grammar chromosome, to provide a grammar that can allow the solution chromosome to express fit solutions effectively. Where the type of change being conducted is a Deterministic oscillating problem, a solution grammar may include potential representations for the previously visited targets.

A consequence of including an extra chromosome for each individual in the population is that the utilisation of computational resources for storing and evaluating the population is effectively doubled. A criticism of explicit memory, laid out in Section 3.3, was that it negated part of the efficiency gain of maintaining and evolving a population across changes in the environment by adding storage space for individuals that were deemed useful enough to be included in memory. The evaluation process was also effectively increased, due to the necessity of evaluating each of the stored individuals whenever a change in the environment occurred. In this case, similar criticism can be made against the incorporation of a second chromosome for each individual. While the grammar chromosome is not explicitly evaluated itself, an extra computational overhead is experienced in storing the extra chromosome and also in conducting the GE mapping process a second time for each individual. Therefore, experiments in this chapter will seek to maintain an equilibrium in the utilisation of computational resources by reducing the size of the population by 50% in the case of the $(GE)^2$ setup.

6.2 Evolving Constants Using a meta-Grammar with Digit Concatenation

This section analyses the use of a meta-Grammar that incorporates the use of pure Digit Concatenation, that is Digit Concatenation without the ability to form expressions. To place perspective on the relative performance of the meta-Grammar, the results in this section will be compared with those in Section 5.4 that used a combination grammar incorporating the Traditional method, Digit Concatenation without the ability to form expressions and Persistent Random Constants.

6.2.1 Problem Domain and Experimental Approach

The constant generation problems tackled are: Finding a Static Constant, Finding Dynamic Real Constants, and the Logistic Difference Equation. This enables the comparison of results of the meta-Grammar approach with results achieved using earlier grammars and to incorporate Markov and Deterministic types of change in the dynamic experiments. The meta-Grammar used for constant generation is provided below.

```
<g> ::= "<SlnCatR> ::=" <catRs>
        "<SlnCat> ::=" <cats>
        "<SlnDigit> ::=" <digit>
<catRs> ::= <catRt> "|" <catRs> | <catRt>
<catRt> ::= "<SlnCat>" | "<SlnCat>"."<SlnCat>"
<cats> ::= <catT> "|" <cats> | <catT>
<catT> ::= "<SlnDigit>"<catT> | "<SlnDigit>"
        | <digit>
<digit> ::= <digitT> "|" <digit> | <digitT>
<digitT> ::= 0|1|2|3|4|5|6|7|8|9
```

A simple example of this meta-Grammar in action is seen in evolving the static target 50. In one such experiment, the meta-Grammar produced the solution grammar displayed below.

```
<SlnCatR> ::= <SlnCat>
<SlnCat> ::= <SlnDigit>0
<SlnDigit> ::= 5
```

This solution grammar then makes it very easy to produce the target, as the only mapping available produces 50. This underlines the strength of the dual structure and use of a meta-Grammar, as it allows the grammar itself to specialise towards the solution.

As the meta-Grammar's performance is being compared to that of the combination grammar of the previous chapter, the grammar is presented below.

```
<exp> ::= <value>
<value> ::= <trad> | <catR> | <persistent>
<op> ::= + | - | / | *
<trad> ::= <trad> <op> <trad> | <tradT>
<tradT> ::= 0 | 1 | 2 | 3 | 4 | 5 | 6 | 7 | 8 | 9
<catR> ::= <cat> <dot> <cat> | <cat>
<cat> ::= <cat> <catT> | <catT>
<catT> ::= 0 | 1 | 2 | 3 | 4 | 5 | 6 | 7 | 8 | 9
<dot> ::= .
<persistent> ::= <persistent> <op> <persistent> | <persistentT>
<persistentT> ::= ''150 randomly generated real constants''
```

6.2.2 Results

The experimental parameters adopted are the same as those outlined in Section 5.3.4 with the exception of population size. In the case of the meta-Grammar runs population sizes of 250 were adopted in order to maintain an equal computational effort with prior experiments. For convenience, Table 6.1 provides a summary of the experimental parameters adopted. A description of the results is given next, followed by a discussion of same.

Table 6.1 Experimental parameters adopted

Parameter	$(GE)^2$	Combo Grammar
Population Size	250	500
Crossover	0.9	0.9
Mutation	0.1	0.1
Selection	Roulette	Roulette
Replacement	25%	25%
Wrapping	8 events	8 events
Codon Size	8 bits	8 bits

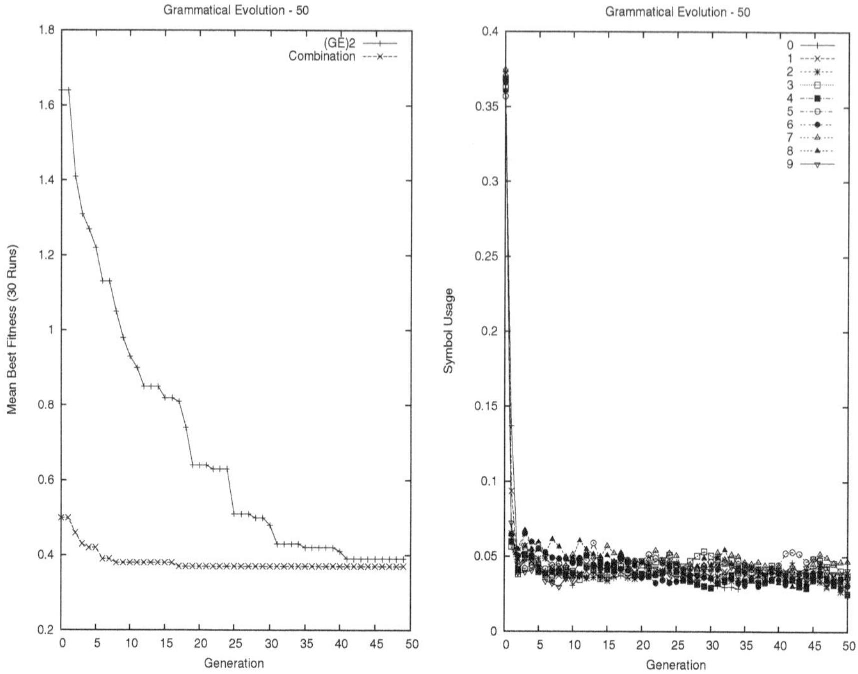

Fig. 6.2 Plot of the mean best fitness values for each constant generation method (left) for the static target 50 and the mean symbol usage at each generation (right) on the static constant problem instance

Finding a Static Constant

The results presented in Figure 6.2 display a comparison of the average best fitness of each of the grammars over the 30 runs. As can be seen the meta-Grammar begins with a poorer fitness in comparison to the combination grammar but quickly evolves a comparable fitness over the 50 generations. A t-test and a bootstrap t-test reveal that there is no significant difference in the results by the final generation. The average best performance of the combination grammar by the final generation was 0.373499 with 6 runs evolving the exact target. In comparison the meta-Grammar produced an average best performance of 0.391333 and reached the target exactly in 14 of the 30 runs.

Finding Dynamic Real Constants

In Figure 6.3 graphs are presented for the dynamic Markov experiments where the set of numbers to be evolved over the course of a run are: 192.47, 71.84, 71.83, 173.59, and 192.47, changing at every 10th generation. Here the Combination grammar again begins with a good fitness. Once generation 10 is reached and the target changes to 71.84, it quickly attains a very good fitness. In these

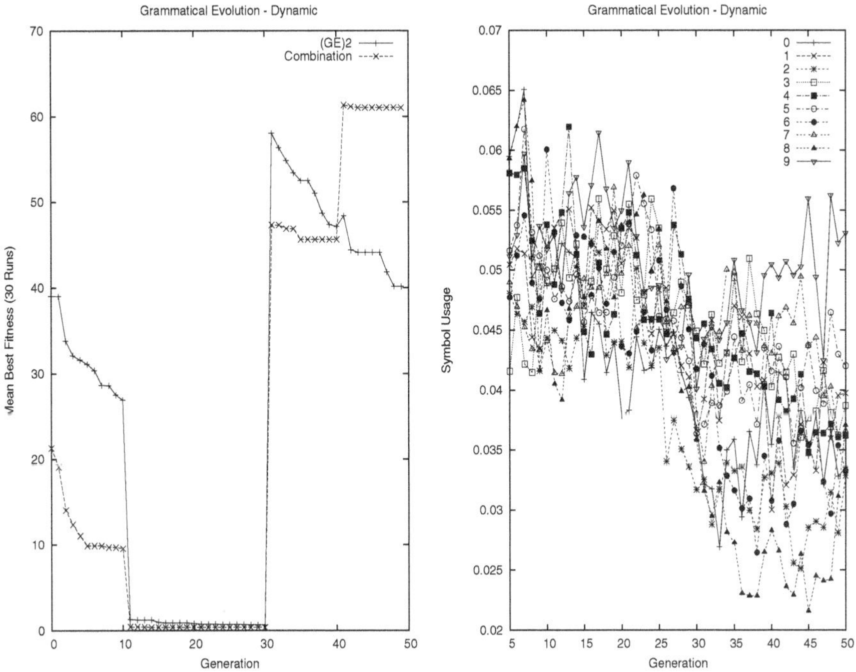

Fig. 6.3 Plot of the mean best fitness values for each constant generation method (left) and the mean symbol usage at each generation (right) on the first dynamic problem instance

experiments, the meta-Grammar method again starts off with a poorer fitness but attains a fitness similar to the combination grammar by generation 10. When the target changes to 173.59, it too experiences a strong deterioration in fitness. However, unlike the combination grammar it takes large leaps in fitness as the generations progress on this target. It then goes on to continue improving fitness when the target again shifts at generation 40 to 192.47.

Results for the Deterministic oscillating non-stationary problem instance are presented in Figure 6.4. In the second instance of this problem where the target oscillates from 192.47 to 71.84 every 10 generations, a similar trend is noticed. For the meta-Grammar method, it again begins with a poorer fitness but quickly catches up with the combination grammar by generation 10. Once the target changes to 192.47, a similar story to the previous dynamic Markovian experiments is observed. The meta-Grammar begins with a poor fitness but quickly evolves more fit individuals over the 10 generations. Interestingly, this trend is emphasised when the target hits 192.74 for the second time where the meta-Grammar begins with a fitness that is worse than at the start the last time, but ends on a fitness that is better than at generation 30.

An interesting aside at this point is to examine the changes in the best-performing solution grammar over the course of an oscillation experiment.

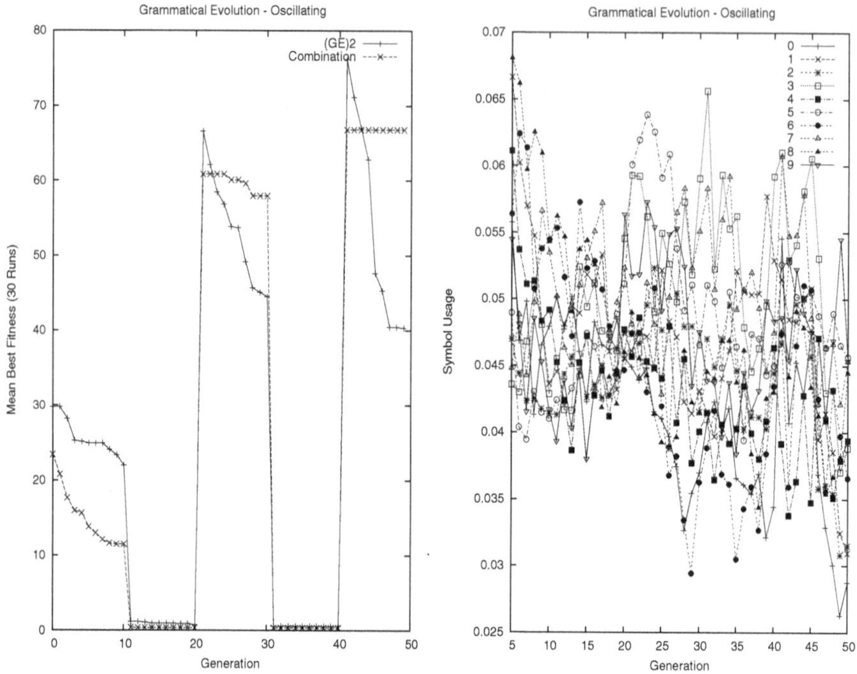

Fig. 6.4 Plot of the mean best fitness values for each constant generation method (left) and the mean symbol usage at each generation (right) on the oscillating dynamic problem instance

In one such experiment, the meta-Grammar was able to produce a solution grammar that helped in closely approximating the 71.84 target. This grammar is described below.

```
<SlnCatR> ::= <SlnCat> | <SlnCat>.<SlnCat>
<SlnCat> ::= <SlnDigit>2
<SlnDigit> ::= 3 | 7
```

A solution yielded by this grammar produced a phenotype of 72.32. The system was then able to maintain this individual in the population when the target swung to 192.47, and recall it again when the target returned to 71.84, demonstrating a population-based memory. A form of representational-based memory is also demonstrated in the grammar below.

```
<SlnCatR> ::= <SlnCat>.<SlnCat>
<SlnCat> ::= <SlnDigit><SlnDigit>
           | <SlnDigit><SlnDigit>
           | <SlnDigit><SlnDigit>6
<SlnDigit> ::= 5 | 6 | 9 | 1 | 7 | 4 | 6
```

Here, a grammar was evolved which allowed the solution chromosome to form phenotypic solutions that were good approximations of both targets. The grammar yielded 196.97 as the solution for target 192.47 and could obtain a good fitness for the target 71.84 by combining the terminals 7 and 1, followed by a pair of floating point digits.

These experiments also saw $(GE)^2$ steadily produce improving fitnesses when the target switched to 192.47 at a better rate than standard GE, highlighting an ability to adapt more easily. Right down to the final generation, better grammars are produced when one best-performing grammar transitioned from:

```
<SlnCatR> ::= <SlnCat>
<SlnCat> ::= <SlnDigit>0
           | <SlnDigit><SlnDigit><SlnDigit>
<SlnDigit> ::= 0 | 1 | 0
```

and a solution of 101 to:

```
<SlnCatR> ::= <SlnCat>.<SlnCat>
<SlnCat> ::= <SlnDigit><SlnDigit><SlnDigit>
<SlnDigit> ::= 0 | 2 | 9 | 1
```

and a solution of 190.021, achieving a good approximation of the target. This is in contrast to the combination-grammar method that generally saw little or no evolution towards the target when it swung to the largest number outside the range of the PRC range.

The Logistic Difference Equation

Here both methods present good fitnesses. Table 6.2 shows average best fitnesses for the different values of α.

Table 6.2 Average best fitness for different values of α for each grammar

α	Combo	$(GE)^2$
3.59	0.000061	0.00032
3.80	0.00045	0.00041
3.84	0.00024	0.0002468

As can be seen in the table meta-Grammar performs well in comparison to the combination grammar with close results in all but $\alpha = 3.59$. Figure 6.2 presents a sample of the results for $\alpha = 3.84$. It once more follows trends seen in the previous experiments, where the meta-Grammar begins with a poorer fitness but rapidly takes the evolutionary steps to reach a fitness similar to the combination grammar.

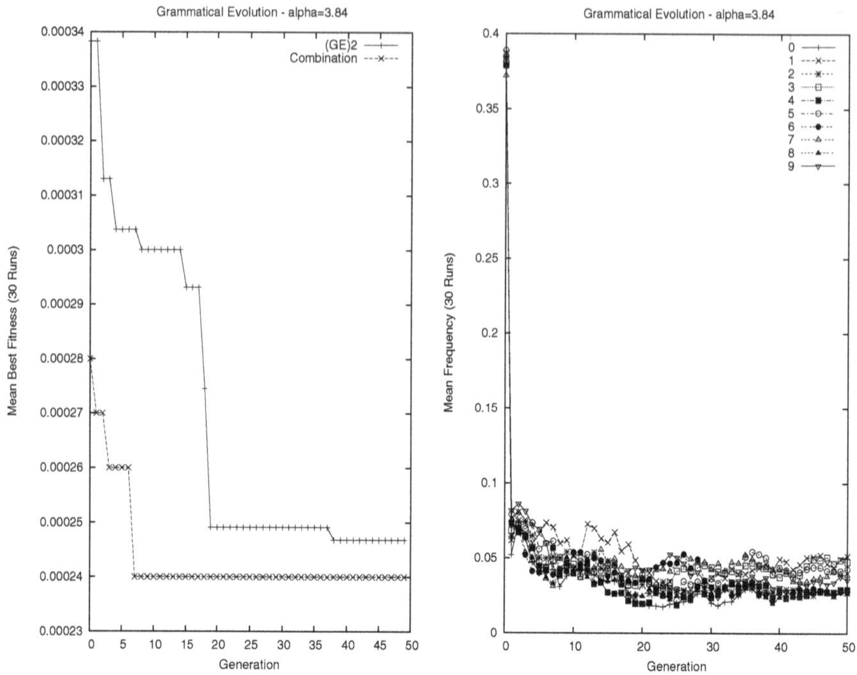

Fig. 6.5 Plot of the mean best fitness on the logistic difference equation problem instance where α=3.84

Discussion

In this section, the combination grammar used in Section 5.4 was compared with a meta-Grammar that adopted the pure Digit Concatenation method, and advantages of the meta-Grammar approach over the combination grammar were seen. These advantages were largely seen in the dynamic experiments where the meta-Grammar has the advantage of being able to evolve to a new target with large evolutionary steps. This is due to its dual-chromosomal structure, where both the grammar and the solution are evolved simultaneously and favourable biases are quickly built upon.

Among the static experiments, the meta-Grammar is able to hold its own with t-tests highlighting there was no significant difference between the methods for evolving the static constant 50. In the logistic equation the meta-Grammar achieved a better fitness in one out of three values for α, and only marginally underperformed at one setting. When considering these results, it should also be recalled that the system was operating with half the population size compared to that of the previous chapter. This demonstrates that efficiency gains can be brought about in the search without increasing utilisation of technical resources.

One of the interesting features of these experiments was the high rate of evolution produced by the meta-Grammar. In all the problem instances, the meta-Grammar began the early generations with a far inferior fitness due to the larger search space presented by the dual chromosome structure. However, over a small number of generations this disadvantage is quickly overcome and fitnesses are attained that are comparable to the combination grammar and its smaller search space, highlighting the utility of the meta-Grammar method.

6.3 Analysis of $(GE)^2$ Using Digit Concatenation with Expressions

Section 6.2 explored the use of a meta-Grammar incorporating pure Digit Concatenation and found that it had advantages over the combination grammar. This section continues the exploration of meta-Grammars by using a meta-Grammar that allows for the creation of expressions incorporating numbers created through Digit Concatenation, utilising the best-explored mechanism from the previous chapter. This method was found to produce significantly better fitnesses over pure Digit Concatenation in Section 5.5.

6.3.1 Experimental Approach

The problems tackled in this section are the same as the previous one. This section, however, uses a meta-Grammar that allows the formation of expressions with the numbers created through Digit Concatenation. Again the same problems are tackled to allow comparisons: finding a static constant and finding dynamic real constants. In this section, the results from experiments using the Digit Concatenation meta-Grammar with expressions are graphed against the results from the previous section, along with the results from a standard GE grammar with expressions and Digit Concatenation from Section 5.5. The new grammar used in this section is shown below.

```
<g> ::= "<SlnExpr> ::= " <catExp>
        "<SlnCat> ::= " <cat>
        "<SlnDigit> ::=" <digits>
        "<SlnOp> ::=" <ops>
<catExp> ::= <catExp> "<SlnOp>" <catExp> | <catRs>
<catRs> ::= <catRt> "|" <catRs> | <catRt>
<catRt> ::= "<SlnCat>" | "<SlnCat>"."<SlnCat>"
<cat> ::= <catT> "|" <cat> | <catT>
<catT> ::= "<SlnDigit>"<catT> | "<SlnDigit>" | <digits>
<digits> ::= <digitT> "|" <digits> | <digitT>
<ops> ::= <op> "|" <ops> | <op>
<op> ::= + | - | * | /
<digitT> ::= 0 | 1 | 2 | 3 | 4 | 5 | 6 | 7 | 8 | 9
```

6.3.2 Results

For every problem instance, the parameters used and number of runs conducted were the same as in Section 6.2.2.

Finding a Static Constant

The results presented in Figure 6.6 display a comparison of the average best fitness of each of the grammars over the 30 runs.

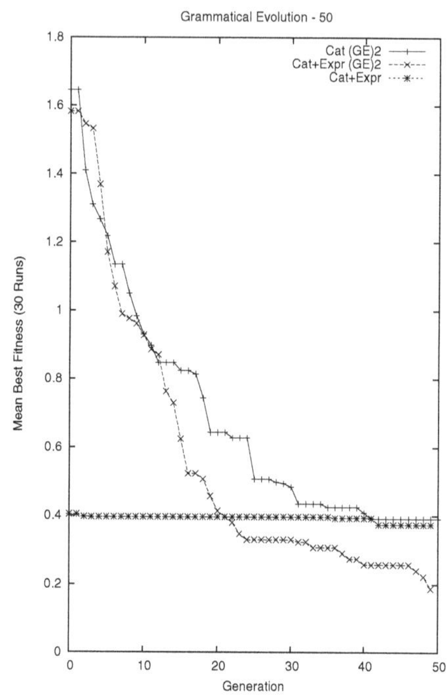

Fig. 6.6 Plot of the mean best fitness values for each constant generation method on the static constant problem instance

The results presented for $(GE)^2$ with the ability to form expressions exhibit similar behaviour to that of the meta-Grammar seen in Section 6.2. However, here an even greater rate of evolution is presented as the generations progress. This strong rate of evolution surpasses standard Cat+Expr average fitness by generation 22, resulting in a final average fitness of 0.187667, versus 0.391333 and 0.373938 for pure Cat $(GE)^2$ and standard Cat+Expr respectively. A t-test and a bootstrap t-test reveal that there is no significant difference in the results by the final generation. In examining the number of runs that evolved the target exactly, $(GE)^2$ with expressions evolved the correct solution 60% of the runs, with Cat $(GE)^2$ on 47% and standard Cat+Expr on 43%.

Fig. 6.7 Plot of the mean best fitness values for each constant generation method on the complex static constant problem instance

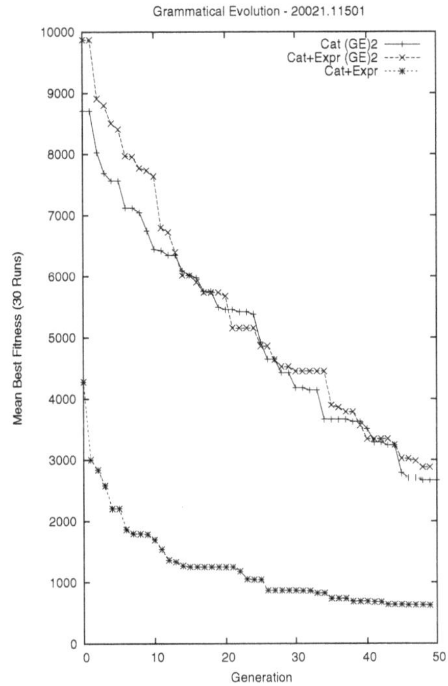

In Figure 6.7 the results for evolving 20021.11501 are presented. Here, neither meta-Grammar method is seen to achieve the level of fitness attained by the standard GE grammar. In this case, both meta-Grammars also displaying a similar rate of evolution with neither having a clear edge by the final generation. The difficulty experienced by the meta-Grammar method is likely down to the larger search space it must navigate, due to the dual chromosome structure, in order to improve its fitness. However, it should be noted that the rate of evolution for the meta-Grammar method, does again, continue at a steady pace right to the final generation. Comparatively, evolution appears to stagnate towards the latter generations of the standard GE Cat+Expr method.

Finding Dynamic Real Constants

Figure 6.8 displays a graph of the average best fitnesses for the dynamic Markov problem where the target changed every 10th generation from 192.47 to 71.84, 71.83, 173.59, and 192.47. Here the standard GE method clearly outperforms the two meta-Grammars with the meta-Grammar method using pure Cat outperforming the Cat+Expr though without any statistical significance under t-test and bootstrap t-tests.

Figure 6.9 displays the graph for the oscillating target problem. Under this experiment, the meta-Grammars perform more competitively against the

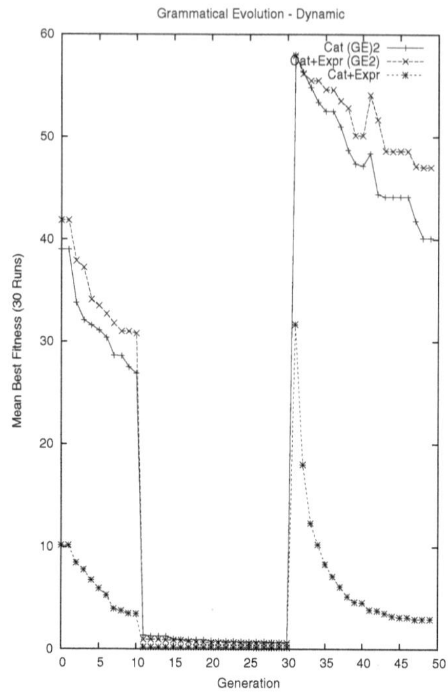

Fig. 6.8 Plot of the mean best fitness values for each constant generation method on the first dynamic problem instance

standard GE approach. For the first generation at each target, the standard GE approach produces better results on average. However, as seen in Section 6.2, this advantage is eroded by $(GE)^2$'s greater rate of evolution. When the target reaches 192.47 for the second time, the Cat+Exp$(GE)^2$ provides the best average performance by the 30th generation. However, when the target returns to 192.47 for the final time, it is the pure Cat$(GE)^2$ that presents the best results that are statistically superior to the Cat+Exp$(GE)^2$ method under t-test and bootstrap t-tests.

Discussion

This section presented a number of interesting results. First, in evolving the static number 50, the Cat+Expr$(GE)^2$ method provided the best results, displaying a strong rate of evolution enabling it to outperform the the standard GE method even with its larger search space and a reduced population size. Evolving 20021.11501 proved to be equally difficult for both meta-Grammars in comparison to standard GE.

In the dynamic experiment, the standard GE method outperforms both the meta-Grammar methods strongly. Here, the smaller search space likely allowed it to gain an upper hand in the transition from 173.59 to 192.47 as it suffers from very little loss of fitness on this change of targets. This performance, however, does not continue in the oscillating experiment where the

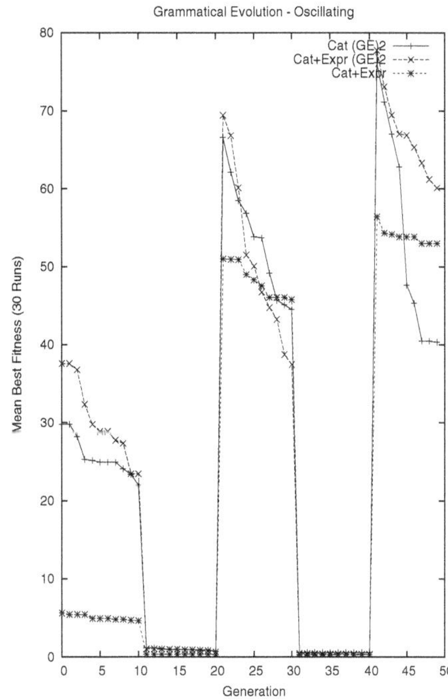

Fig. 6.9 Plot of the mean best fitness values for each constant generation method on the oscillating dynamic problem instance

standard GE approach suffers from a weak rate of evolution when the target transitions to 192.47. The meta-Grammar approach, on the other hand, again displays a steady rate of evolution, with one of the two meta-Grammar methods performing better by the end of each of the final transitions to 192.47. Interestingly, the meta-Grammar that uses the pure Digit Concatenation method for constant creation performed better than the method which incorporated expressions with a statistical significance.

6.4 Conclusions

This chapter uncovered some of the potential of the meta-Grammar approach and $(GE)^2$ for dynamic environments. The addition of the extra layer of adaptability allowed $(GE)^2$ to take large evolutionary steps when a change in the target occurred. Across all problems, a high rate of evolution was observed right through to the latter generations. Evidence was also observed, though anecdotal, of a representational-based memory forming in the solution-grammar chromosome for the Deterministic oscillating problem.

In examining the basic functionality of $(GE)^2$, it was seen to perform comparatively well on the static experiments, presenting a higher ratio of exact matches for the target 50 though with a slightly inferior performance on the large number. For the dynamic experiments, $(GE)^2$ demonstrated a higher rate of adaptability in most experiments. In the Deterministic oscillating problems, $(GE)^2$ was able to recall previously successful grammars when a target was encountered again, offering evidence of an implicit memory.

It should also be highlighted that the addition of the meta-Grammar represents an efficient improvement in GE's search capability. Even though the population size was reduced by 50%, $(GE)^2$ was able to perform comparatively with standard GE and present improvements in performance on dynamic problems. This is particularly significant as the 50% reduction implies that there is half the feedback back to the population from fitness evaluations while the search space itself has grown with the addition of the extra chromosome.

Chapter 7
Controlled Static Trading with GE

Previous chapters in this book have examined the evolution of constants with the aim of identifying the most effective method of constant creation and adaptation. These experiments were conducted, for the most part, unencumbered by the added complexity of having to evolve the constants in line with function terminals. This enabled the development of a clear picture as to the behaviour and performance of different constant creation methods. In this chapter, GE is embedded in a more complex environment – that of financial markets. The aim here is to generate trading rules to optimally navigate a price series. However, as before, controlled and static experiments are initially conducted so as to determine the efficacy of the method adopted and to gain insight into the approach's behaviour without the complexity of a shifting-fitness landscape.

This chapter also provides a background on the pervading theories on financial markets, technical analysis, and conducts a brief review of the work to date using GA/GP paradigms in the field of trading. The trading model adopted in this book is introduced and experiments are conducted over static artificial market scenarios in order to demonstrate functionality and gain insight into the system's behaviour under different scenarios. The aim is to identify if the system is capable of evolving optimal rules on a noiseless, controlled, static, price series. These experiments will also contribute to determining system parameters for when it is applied to dynamic environments.

The next section provides an introduction to the problem domain. This is followed by Section 7.2 which gives a background on modern financial theory and highlights some of the issues faced in trading. Section 7.3 introduces technical analysis and Section 7.4 conducts a review of studies where EC has been applied to trading. Section 7.5 describes the trading model adopted used. A section on initial experimentation is next and Section 7.7 closes the chapter with conclusions.

7.1 Introduction

A number of dynamic problems have been examined thus far in this book that have focused on two types of change: *Markov* and *Deterministic*. In this

I. Dempsey et al.: Foundations in Gramm. Evol. for Dyn. Environ., SCI 194, pp. 121–140.
springerlink.com

chapter, an initial examination of the third type of change outlined in Chapter 3, *Complex* change, is conducted. De Jong [52] described this type of change as one where new regions of fitness may emerge that previously contained no optima and that competitive problems embody such changes. Under these conditions, the different strategies used by the competing participants give rise to a complex aggregate behaviour that is difficult to predict or extract information from. This is because, as the competition proceeds in time, the participants will adapt and change their approaches to gain an edge over the other parties.

Trojanowski and Michalewicz [215] added that complex changes are dependent upon prior state and changes, but are the product of relationships too complex to predict. Again, this points towards competitive problems where entities are continually changing their modus operandi in order to improve their performance relative to others.

Trading on financial markets satisfies such criteria. The financial markets of the world are dynamic in nature, with a vast number of participants competing against one another in the pursuit of profits, across varying time horizons, under different constraints. Therefore in applying GE to financial markets a useful test-bed is created for exploring different approaches to navigating dynamic environments that can be replicated in other domains. The next section will provide a background on modern financial theory, shedding light on why they present a Complex type of change.

7.2 Modern Financial Theory

Vast sums of money flow through the markets of the world on a daily basis. With this in mind it is not surprising then, that the domain has attracted an extensive amount of research with the aim of understanding or gaining better insights into the mechanics of price formation and market dynamics.

7.2.1 *Efficient Market Hypothesis*

Bachelier first drew the link between financial markets and Brownian motion in 1900 [8]. In 1953, Kendall [104] followed in observing that prices on the London stock exchange essentially followed a random walk where a price's direction was as likely to go up as it was to go down, irrespective of past prices. This then led to further research which asked the question: what drives price formation? Further study in the 1960s by Eugene Fama then led to the proposal of the Efficient Market Hypothesis (EMH) [64]. This hypotheses states that the market price for a security reflects all information available at the current point in time. Much of modern financial theory is predicated upon the EMH. Since being proposed, the EMH has been the cornerstone behind theories such as the Capital Asset Pricing Model and the Black-Scholes option pricing model. This is because these theories make assumptions on market efficiency, rational investors, and normally distributed returns. Three forms of the EMH are defined that aim to take into account the different rates at which information is assimilated to drive price formation. These are:

i. **Weak Form EMH:** This is where all historic price and volume information is already factored into the current price.
ii. **Semi-Strong Form EMH:** This form states that all publicly available information is already factored into the current price. Such information includes all fundamental information on the security, and price and volume data.
iii. **Strong Form EMH:** Here, all information, whether it be public or privately held insider information, as well as price and volume data, are all incorporated into the current price of the security.

The weak form implies that analysis of historical price and volume data will not yield an edge over the market in terms of performance. While semi-strong and strong imply that no advantage can be gained from expanding the scope of information used and examining the fundamental or private data of the security. Such implications render the use of technical analysis, and in the stronger forms any kind of fundamental analysis, useless as tools for researching the potential performance or the directionality of a security's movement with an aim to beating the market. Therefore, under the circumstances of an efficient market, securities are fairly priced in the light of all information available. If this is the case, then investing is an expected zero-NPV (Net Present Value) transaction without taking into account transaction costs.

It would appear, however, that this perspective is at odds with much of the financial industry where vast sums of money are spent developing quantitative models and conducting both technical and fundamental analysis. Consistent excess returns over market benchmarks are also demonstrated by many hedge funds [47], along with noted investors such as Peter Lynch and Warren Buffet, who have generated consistent excessive returns over long periods of time. To quote a notable investor, George Soros, on the EMH, *"this (EMH) interpretation of the way financial markets operate is severely distorted... It may seem strange that a patently false theory should gain such widespread acceptance"* (1994). However, in order to make the EMH operationally testable it must be structured jointly with an appropriate asset pricing model that takes into account the investors risk profile [122]. Indeed when structuring the model in such a way it becomes difficult to refute the EMH specifically as either component of the joint hypothesis may in fact be leading to the inconsistency.

7.2.2 Behavioural Finance

Fundamental to the EMH is that all information is incorporated instantaneously into a security's price. It assumes that all market participants are rational beings and make optimal decisions on the information they receive.

However, real world infrastructural limitations can obscure the dissemination of this information resulting in different parties being privy to different items of information.

Evidence from the field of Behavioural Finance suggests that humans, who constitute the population of market participants, are subject to a number of irrational shortcomings when making decisions in uncertain environments. The result of this is that inefficiencies are created through actions due to such decisions. Shefrin identifies three main areas of Behavioural Finance [200].

i. **Heuristics:** This includes risk aversion or risk seeking behaviour [98], or overconfidence [187].
ii. **Framing:** This refers to how different decisions can be arrived at depending on how the choice is described or its context[221].
iii. **Market Inefficiencies:** This is where there are explanations for observed market inefficiencies that run counter to the rational expectations of the EMH [213].

Behavioural Finance essentially replaces some of the ideal assumptions adopted in the EMH. In their place, Behavioural Finance borrows theories from the field of Psychology to model the decisions investors (humans), make under uncertainty where rewards and losses are in the balance.

While Behavioural Finance is receiving an increasing amount of traction, it is not without a certain element of controversy. Subscribers of the EMH argue that it is largely an empirical list of counter examples with little in the way of a solid unifying principle.

7.2.3 Adaptive Market Hypothesis

Another view of the market is given by Lo where he attempts to reconcile the conflicts between the EMH and Behavioural Finance by suggesting that the market is itself an evolutionary process [124, 125]. Lo describes his theory as the Adaptive Market Hypothesis (AMH). In the AMH, Lo suggests that the market is made up of an ecology of investors that is permanently evolving with the evolutionary pressures of diversity and competition for resources all playing roles in this process. It is these pressures that then determine the efficiency of markets.

In reconciling the differences between Behavioural Finance and the EMH, Lo argues that *"...behaviour biases are simply heuristics that have been taken out of context, not necessarily counterexamples to rationality"*. He also argues that instead of maximising expected utility and having rational expectations, investors are organisms that have evolved to maximise their chances of survival in the current context. Stemming from this, the AMH presents five implications.

i. The equity risk premium is not constant through time but varies according to the recent path of the stock market and the demographics of investors during that path.
ii. Asset allocation can add value by exploiting the markets path dependence as well as systematic changes in behaviour.
iii. All investment products tend to experience cycles of superior and inferior performance.
iv. Market efficiency is not an all-or-none condition but is a characteristic that varies continuously over time across markets.
v. Individual and institutional risk preferences are not likely to be stable over time.

Key to these implications is that the market participants are not absolutely rational but instead engage *satisficing*, a type of bounded rationality identified by Simon [203]. In the AMH, these boundaries are established through the experience of the investor as reinforcement is received through interactions with the market.

7.2.4 Summary

While the markets of the world are remarkably efficient, contributions from Behavioural Finance and elsewhere demonstrate that trends do exist and market anomalies can occur through investor overconfidence, exuberance, errors in judgment, or infrastructural problems. The fortunes amassed by individuals and institutions in the real world suggest that new information extraction techniques can yield an edge in the competition for returns, though over time such techniques may be employed by a sufficient number of other participants to render it unprofitable. Efficiency is not an inbuilt market property but a consequence of the interpretation, by investors, of the information to which they are exposed.

Consequently, the financial markets meet the criteria for a dynamic complex problem where competition among investors generally creates efficient prices, while opportunities for benefiting from trends and anomalies can also provide investors with the means to outperform market benchmarks.

7.3 Technical Analysis

Technical analysis can be defined as an attempt to identify regularities (or irregularities) in the time series of price and volume information from a financial market [123]. If it is accepted that prices can move in trends and patterns repeat themselves [138], then rules may be mined that underlie price behaviour. Given their environment they will necessarily be dynamic. Indeed, studies have been conducted that give credence to the use of technical rules [34, 99]. With regards to their application in the real world, Taylor and Allen [212] found that 90% of institutions dealing in foreign exchange in London place some weight on information obtained through technical analysis.

7.3.1 Technical Indicators

The development of trading rules based on current and historic market price information has a long history [35]. The process entails the selection of one or more technical indicators and the development of a trading system based on these indicators. These indicators are formed from various combinations of current and historic price information. Although there are potentially an infinite number of such indicators, the financial literature suggests that certain indicators are widely used by investors [34, 138, 176]. Four groupings of indicators are given prominence in prior literature. They are:

i. Moving average indicators,
ii. Momentum indicators,
iii. Trading range indicators and
iv. Oscillators.

The first two indicators can be used to produce signals in trending securities. Moving averages smooth historic price series to give greater insight into an underlying trend. Buy or sell signals can be generated when a trend reverses. Moving averages, however, can suffer from a lag as signals are generated after a trend has turned. Momentum indicators identify how strong a trend currently is. When a trend weakens, a position is unwound or built up to fade the trend. Trading range indicators identify a range in which the security trades. When this range is broken, action is taken. Oscillators aim to identify the state of the security by determining if it is over-bought or over-sold.

For the purposes of these experiments, the focus will be on moving average indicators. A long moving average and a short moving average can be combined to produce an oscillating signal – the Moving Average Convergence Divergence (MACD) indicator. As already stated, this is a trend identifying indicator. When the short moving average crosses above the long a buy signal is generated and conversely a sell signal is generated when the long moving average crosses above the short.

7.4 EC and Technical Trading

Over the past ten years, a number of studies have been completed by researchers investigating the potential of applying EC paradigms to trading. Table 7.1 provides a summary of these approaches. All approaches listed employ some form of technical analysis in generating the buy and sell signals that are input to their trading systems underlining the significance attached to the technical methodology.

With regard to the results, the studies present mixed findings. Chen and Yeh approached their work with a view to determining the efficiency of stock markets and saw inferior results to a benchmark buy-and-hold strategy. Neely et al. saw superior risk-adjusted returns in their work on evolving rules for trading currencies on the foreign-exchange markets. Allen and Karjalainen

Table 7.1 Summary of research in applying GA/GP paradigms to trading

Authors	Type	Desc.	Results	Dynamic	Trans. Costs
Chen & Yeh [40]	GP	Evol math exprs to forcast return	Inferior to buy & hold	No	0
Neely, Weller & Dittmar [139]	GP	Evol tech rules for ForEx	Superior, risk adjusted	No	0.1%
Allen & Karjalainen [2]	GP	Evol daily tech rules for mkt indices	Inferior to buy & hold	No	0.25%
Li & Tsang [120]	GP	Evol decision trees	No comparison	No	0
Becker & Seshadri [11]	GP	Evol monthly tech rules for mkt indices	Superior to buy & hold	No	0.5%
Potvin, Soriano & Vallee [174]	GP	Evol daily tech rules for stocks	Inferior to buy & hold	No	0
Kwon & Moon [114]	GA/ANN ensemble	Evol ANNs using tech indicators	Superior in most cases	No	0.3%
Schoreels & Garibaldi [197]	GA	Evol params for tech rules	No comparison	Yes	0
Dempster & Jones [57]	GA	Evol params & structure for ForEx tech rules	Mixed	Yes	0.1%

saw inferior results in their contribution to stock market analysis again, but Becker and Seshadri saw strong returns in their approach.

The level of transaction costs levied on trading also varies across the works examined here.

7.4.1 Dynamic EC and Technical Trading

Similar comments may be attributed to the application of dynamic EC paradigms as have been made previously in this book, in that a significant body of work exists in the application of static GA and GP approaches to financial analysis. However, with regard to dynamic approaches the available literature is sparse as seen in Table 7.1. The financial markets are dynamic in nature – continuously changing and responding to new information and the actions of its participants. The employment of a certain static model may yield positive returns for a period of time. However, as changes are compounded, the static model will grow obsolete. As the possibility exists for these changes to occur quickly, subscribing to a static approach can translate to a fast accumulation of losses. It is necessary, therefore, to adapt the model or indeed allow for the capability of new models to be generated to take advantage of new market states.

Schoreels and Garibaldi [197] present a study where a GA is utilised to evolve parameters to a set of technical rules. A number of experiments are conducted to find an optimal set of retraining parameters, as the population is adapted across the time series. Random immigrants are used to actively maintain diversity within the population. In order to overcome problems with measuring performance in the dynamic environment, Schoreels and Garibaldi adopt the area under the total assets chart as a fitness measure. However, having only conducted five runs for each set of parameters and not accounting for transaction costs, the results do not carry strong statistical relevance.

Dempster and Jones [57] employ a GA-type paradigm to adapt a population of technical trading rules for intra-day foreign-exchange trading. Fixed-length strings are used that include all potential indicators that may be adopted. Similar then to Ohkura [165, 166], areas of the string are activated to form variable-length phenotypic rules. As highlighted in Section 3.3.2, this then allows for the potential of implicit memory to form in inactive regions and also for the occurrence of neutral mutation. However, an effect of this was a reduced search space as there was a limited number of possible strategies that could be chosen. As a result, the population size was then set to 100 individuals. In doing so, the authors are also essentially bounding the model structure and not allowing for its discovery through evolution, limiting the potential scope. In this study, Dempster and Jones preemptively maintain population diversity. They do this by only allowing genotypes that have been altered by genetic operations to replace the weakest members of the population, if the resulting genotype is unique within the population. Convergence within the population is also monitored, whereby if a user-defined number of the best individuals' fitnesses and the maximum fitness both change by no more than 1% evolution is stopped. A user-defined minimum number of generations must first be observed. A weak set of results are reported however, where the executed runs were largely loss-making with the exception of the best run, which generated modest but statistically significant excess returns.

In applying an evolutionary system to financial markets, lessons are learned that can be transferred to other dynamic domains where Complex changes occur. Financial markets represent a difficult and useful test-bed for the application of GP-type paradigms where model discovery and adaptation is necessary to produce positive results. A successful system in this domain must overcome problems associated with dynamic environments generally, leading to insights that are applicable elsewhere.

7.5 Trading Strategy

In the trading model developed here, the rules evolved by GE using the grammar below [53], generate a signal that can result in one of three actions being taken by the model: Buy, Sell, or Do Nothing. Through the use of GE and the BNF grammar, the rules produced by the system are human readable and easily understood. The facility is also available to incorporate certain domain knowledge into the grammar. In the grammar below, the `<expressions>` non-terminal produces the signal through expressions using the `MA` terminal, which is a moving average function. GE may then combine these `MA` terminals to form the MACD indicators described in Section 7.3.1, as evolution sees fit.

```
<expressions>::=<expressions><op><exp>|<exp>
<exp>::=MA( <intExp> )|<intExp>
<op>::=-|+|*|/
<intExp>::=<int><op><int>|<int>
<int>::=<int><digit>|<digit>
<digit>::= 1 | 2 | 3 | 4 | 5 | 6 | 7 | 8 | 9 | 0
```

As can be seen from the above BNF grammar, Digit Concatenation is used with the ability to form expressions as the method of choice for constant creation and adaptation. As this grammar is designed with the MACD oscillator in mind, price series exhibiting trend and reversal behaviour will be ideally suited to this grammar. GE should search the space of different sized moving windows to identify optimal points at which to buy or close positions while at the same time not being sensitive to noise or unprofitable short term movements. For example given a trending price series with a strong signal to noise ratio a sensitive MACD oscillator would be useful:

```
MA(5) - MA(10)
```

while under more noisy conditions the time frames would need to be longer in order to identify the more meaningful underlying trend:

```
MA(20) - MA(50)
```

Though these rules are clean MACD oscillators the scope also exists within this grammar to evolve rules such as:

```
MA(537)*MA(2252-7)-MA(261)-MA(22)*MA(913+82)
```

7.5.1 Variable Position Trading

In prior studies that applied GE to evolve financial trading systems [22, 17, 152, 150, 149], the entry strategy for each trade was to invest a constant monetary amount on the production of a Buy or Sell signal. The relative strength of the buy or sell signal was not considered. Here in contrast, the trading system adopts a more complex entry strategy, and a *variable-size* investment is made, depending on the strength of the trading signal. The stronger the signal the greater the amount invested, subject to a maximum investment amount of \$1,000 (arbitrary system parameter). Therefore, the amount invested for each signal is:

$$Amount\ invested = \frac{Size\ of\ trading\ signal}{Maximum\ trading\ signal} * MAX\_AMNT \qquad (7.1)$$

Signals received from individuals will oscillate around a pivot point of zero. Signals greater than zero constitute a buy signal. To allow the system to decide how much to invest in a given trade, the maximum size of a trading signal must be determined. Initially, the size of the maximum signal is set as being the size of the first buy signal generated by the system. If a signal is subsequently generated that is stronger than this, the maximum trading signal is reset to the new amount, weaker signals invest a smaller amount in the same ratio as their signal to the maximum.

If the sum to be invested is greater than the cash available, the model will invest the cash available less the costs of the transaction. Upon receipt of a sell signal (one less than zero) all positions are closed.

7.5.2 Return Calculation

The total return in this model is a combination of its generated return from market activity and the interest gained from the risk-free cash position. For the experiments in this chapter, US interest rates for the period 1/1/1991 to 1/10/1992 were used, affording an accurate reflection in the cash position. Transaction costs in this model are based on the cost structure used by online trading houses, where flat fees are incurred for the opening and closing of positions: a conservative \$10 fee is charged upon entry and exit. When the transaction costs represent a certain percentage of the sum to be invested, the investment becomes unfeasible. Therefore, the model is made aware of the transaction costs. When the entry and exit costs amount to 20% of the sum to be invested the model will not take the position, effectively ruling out very small investments arising from weak trading signals or frequent trading due to sensitivity in horizontal markets. This represents the Do Nothing stance, where the model will hold all current positions.

Thus, the pure return over the training period was chosen as the fitness measure. It should be noted that while the pure return of the system is useful in forming the competitive infrastructure necessary for evolution to take place, this metric is not necessarily as useful analysing the system as an investment strategy. This is because when measuring the performance of an investment strategy trading returns need to be differentiated from market returns as well as adjusting the returns for risk. For simplicity the total return of the system is compared to the buy-and-hold return of the index in the dynamic experiments going forward. This comparison suffers from problems already outlined in evaluating dynamic systems as it represents a single snapshot of the performance of the system. For example the S&P 500 data set adopted in this study embarks on a sustained upward trend for the majority of the period under investigation making it very difficult, at the end of the period, for the trading system to outperform the benchmark. However if extra data was also used where the S&P 500 drops in value the system would be given the opportunity to close out of its positions at the peak allowing it to demonstrate returns better than the benchmark.

7.6 Static Artificial Data

These experiments used artificial data to examine the performance of the system under specific circumstances. Five cases were tested.

i. A linear upward trending series.
ii. A linear downward trending series.
iii. A horizontally trending series.
iv. A series embarking on a linear upward trend for half of the data series and linear downward trend for the second half.
v. An oscillating series that trends up and down at equal rates.

No live or dynamic element was included in these experiments. Here the goals are:

i. To determine whether GE is functional under this domain, in effect to demonstrate that sensible rules are being evolved as a result of training on the given data.
ii. The experiments are taking place without any noise, under static conditions, with controlled artificial scenarios. Under such conditions, GE should be expected to evolve optimal or near optimal rules.
iii. Ensure that GE through the use of its BNF grammar that incorporates the moving average indicator, is capable of identifying the trends that are artificially placed in the series.
iv. If the trends are identified, how accurate are the rules that are evolved in identifying trend transitions.

Essentially, these experiments are designed to demonstrate the fundamental functionality of the system. If it cannot be shown that the system can evolve models that navigate a static controlled price series optimally then trying to determine its behaviour in a noisy dynamic environment becomes much more difficult. Also, through analysis of the results, insights will be provided into the behaviour of the system that will then help in determining system parameters for applying the system to a dynamic environment.

As these experiments take place over a static data set, the population is just exposed to the same data over training. Because of this, some of the weaknesses of static training should be observed, such as converged populations and brittle, over-fit solutions.

With this in mind, a range of information from the run was logged. The number of individuals that had equivalent performance to the best performer in the population was logged in order to measure the convergence within the population. An individual was deemed equivalent to the best-performing individual if the difference in fitness was less than 0.009. The average fitness of the entire population along with the fitness of the best individual was logged, where fitness was the average rate of return per month. The phenotypic value of the best individual for each generation was also recorded. Each run was executed over 365 days of data, with a 75-day buffer at the beginning of the series to provide scope for the moving average technical indicators and the system was only allowed adopt long positions. Table 7.2 provides a summary of the system parameters adopted.

Table 7.2 Experimental parameters adopted

Parameter	Setting
Population Size	500
Crossover	0.9
Mutation	0.1
Selection	Roulette
Replacement	25%
Wrapping	8 events
Codon Size	8 bits
Bank	$10,000
Interest	US Treasury
Max Order Size	$1,000

7.6.1 Up Series

Here, a linear upward trending data set was used. The goal of this experiment was to determine if the system would generate trading signals that would lead it to take maximum-size positions on consecutive days from the outset. Because the system adopts variable-sized positions, the ideal trading rule would have to generate buy signals equal to, or greater than, the previous day to allow for the maximum investment size. Figure 7.1 describes the average

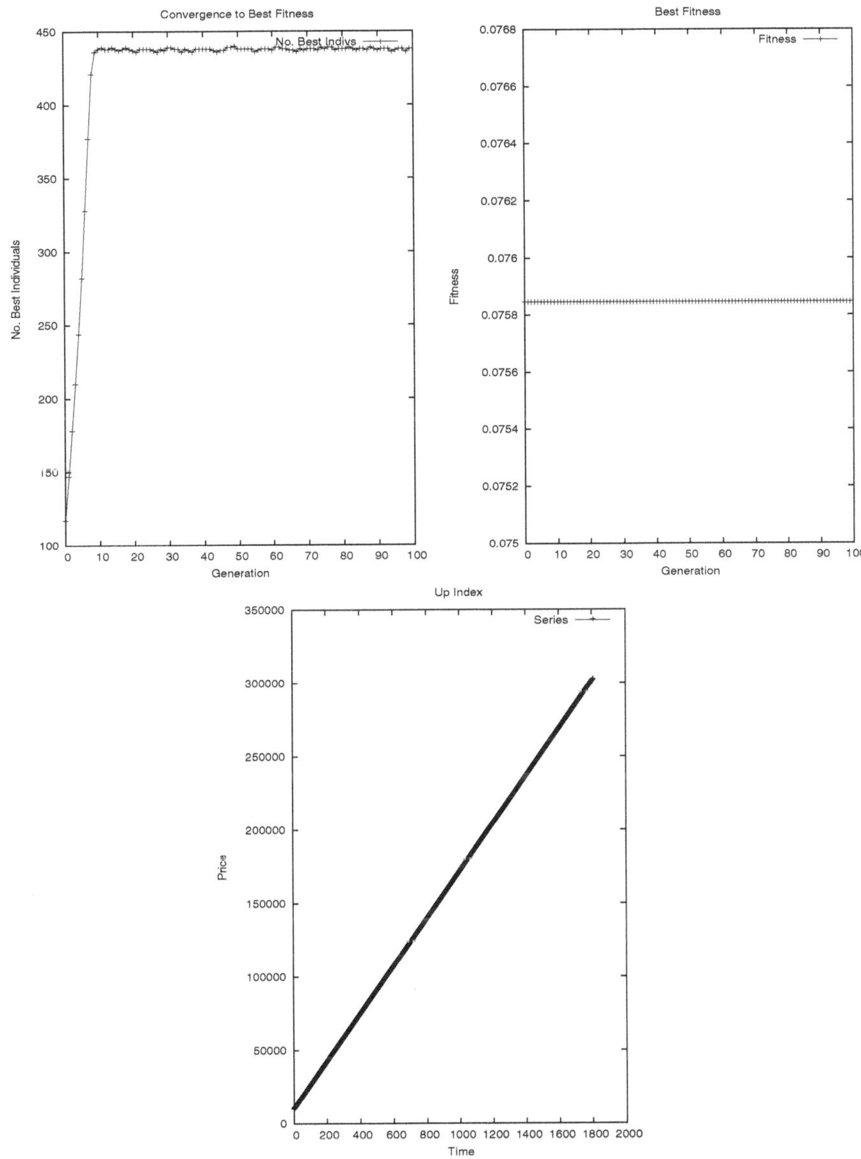

Fig. 7.1 Number of individuals with fitness equivalent to the best individual, upward trending series (top left), average best fitness (average monthly percentage return), upward trending series (top right), and the experiments price series (bottom)

number of individuals who obtained a fitness equivalent to the best performer, whose fitness is optimal at each generation, the average best fitness at each generation, and the artificial price series used.

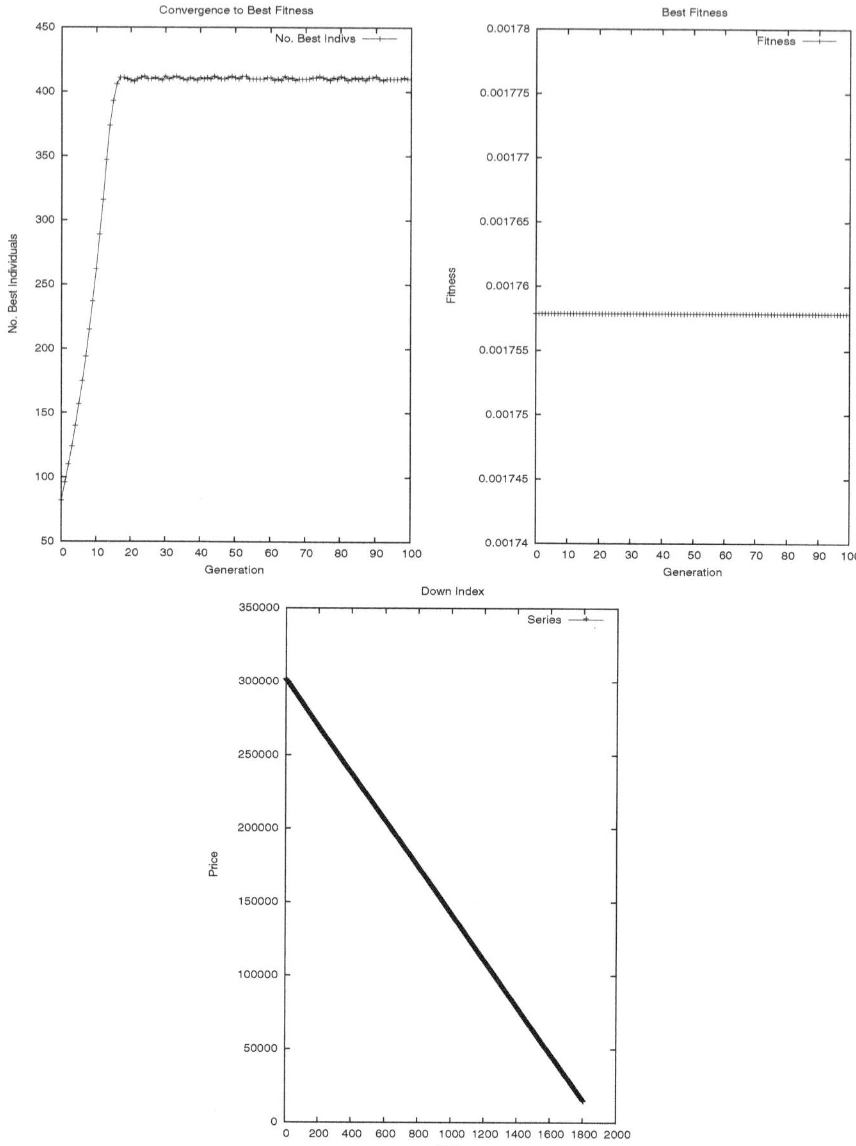

Fig. 7.2 Number of individuals with fitness equivalent to the best individual, downward trending series (top left), average best fitness (average monthly percentage return), downward trending series (top right), and the experiments price series (bottom)

In examining Figure 7.1 it can be seen that the population essentially converged before the 10th generation. The average best fitness is seen to flat line for each run. This is due to the fact that a rule to generate the same signal every day is easily obtained as it could be as simple as the constant "1".

7.6.2 Down Series

The ideal solution to a linearly down trending data set is to generate sell or do nothing signals. This enables the system to maintain a cash position and accumulate compound interest on its funds. Figure 7.2 describes the average number of individuals who obtained a fitness equivalent to the best performer at each generation, the average best fitness at each generation, and the artificial price series used.

These results are very similar to the upward trending results. In this case, the generation of a signal less than, or equal to, zero will deliver the best results.

7.6.3 Even Series

Like the downward trending data set, the ideal solution here is to again generate sell or do nothing signals. In this case, however, the negative reinforcement signal is not quite as strong, as the only cost involved in entering the market is the trading fees and a reduced exposure to interest-rate returns as funds invested do not accrue interest. Figure 7.3 describes the average number of individuals who obtained a fitness equivalent to the best performer at each generation, the average best fitness at each generation where this fitness is optimal, along with the price series used. It can be noted that the return is the same as that in the previous experiment. This is because the best performers in both cases maintain cash positions and so are exposed to the same interest rate returns.

7.6.4 Up Down Series

This data set requires the generation of a more sophisticated rule. For half of this data set the series trends linearly upward and on day 220 it trends downward at the same rate at which it increased. The ideal solution here is to generate a rule that will buy in the same manner as that in Section 7.6.1, but close all postions when it recognises that the series has turned. Figure 7.4 describes the average number of individuals who obtained a fitness equivalent to the best performer at each generation, and the average best fitness at each generation, along with the artificial price series used.

It may be noticed in Figure 7.4, that there is a longer climb than drop in the price series. This extra space is due to the 75-day buffer for rules at the start of the series. An optimum investment strategy here would invest the maximum amount of its funds each day, initially generating an average price of 49768.5 and sell at the peak, 81291.7. This would generate excess returns of 64% over the period of the data set. Figure 7.4, displays the average return per month of the best individuals, with the final generations generating monthly

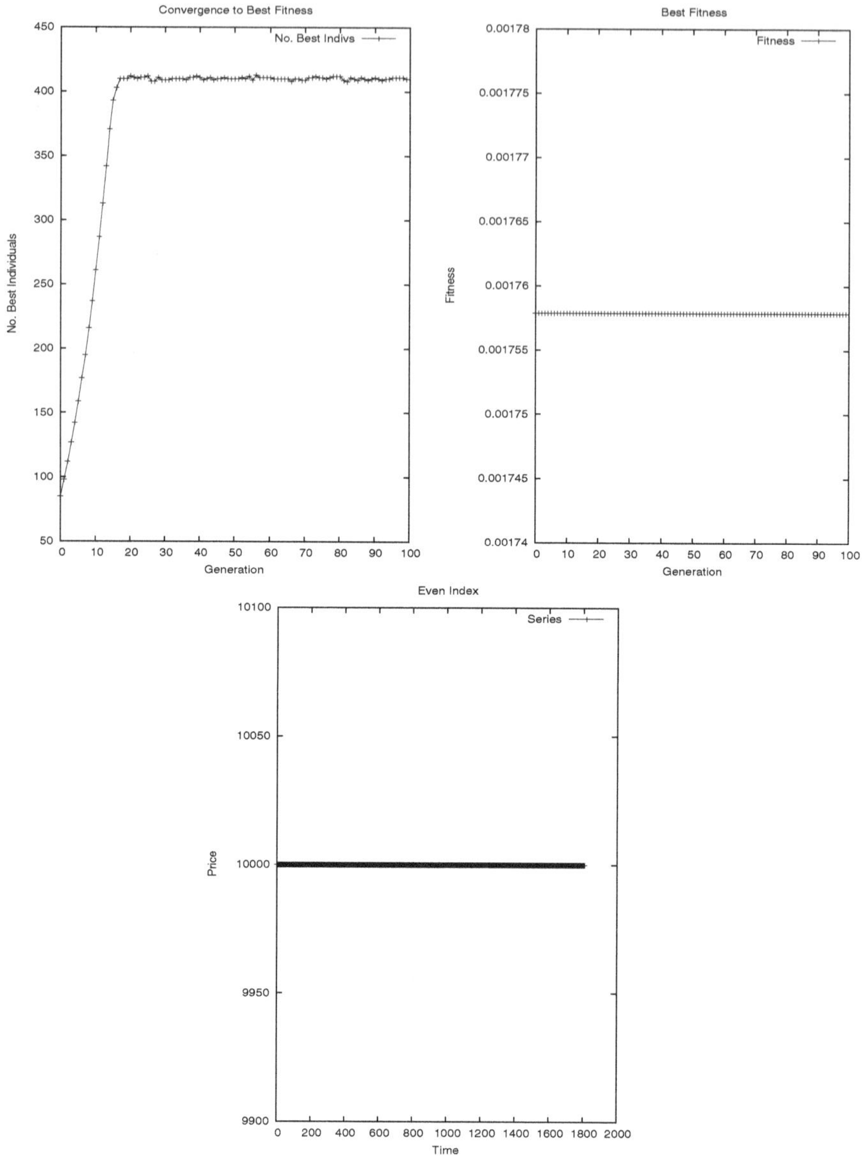

Fig. 7.3 Number of individuals with fitness equivalent to the best individual, horizontal trending series (top left), average best fitness (average monthly percentage return), horizontal trending series (top right), and the experiments price series (bottom)

returns of almost 2.7%, indicating that all runs gave a return greater than 64%. This means they sold within a day or two of the peak and then benefited from interest in a cash position after the peak. Convergence in this problem

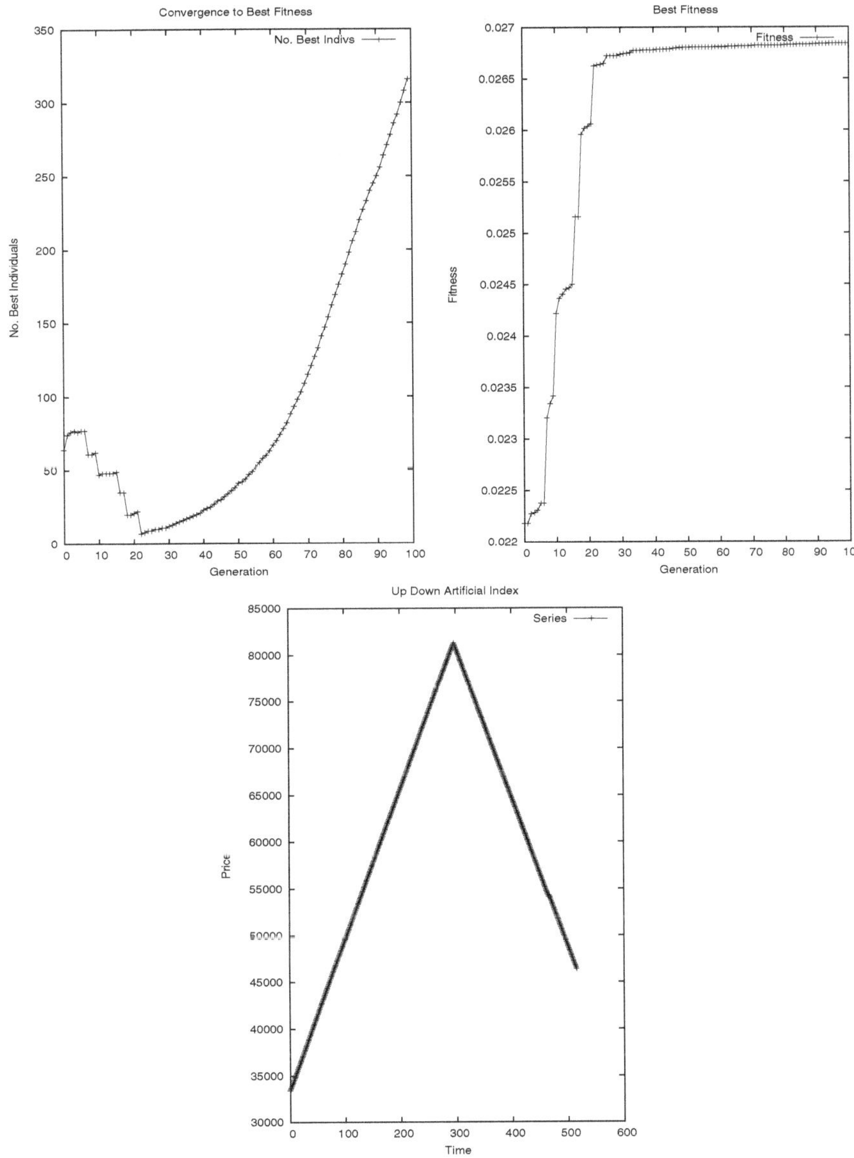

Fig. 7.4 Number of individuals with fitness equivalent to the best individual, up down trending series (top left), average best fitness (average monthly percentage return), up down trending series (top right), and the experiments price series (bottom)

occurs later than in earlier problems with slight improvements to fitness right up until the end. Interestingly, the dramatic increase in fitness for the first 20 generations is matched by a corresponding drop in the number of individuals

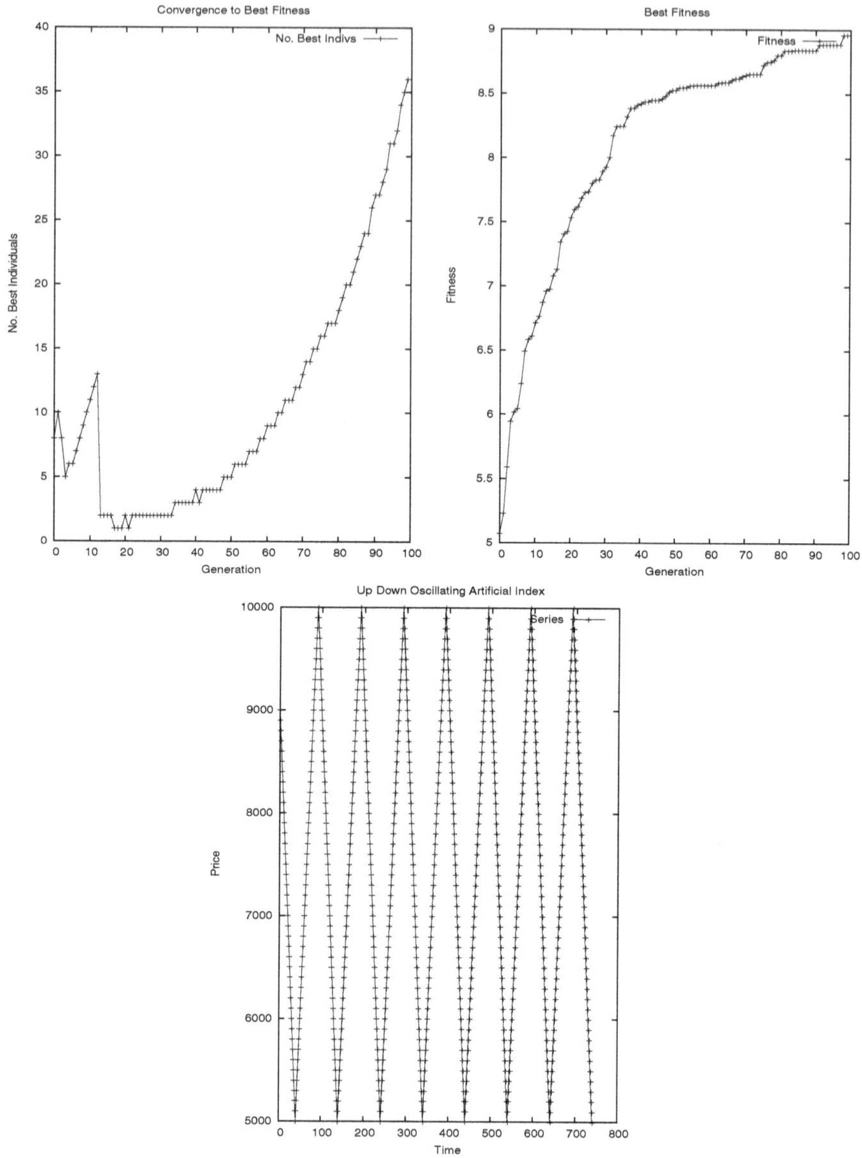

Fig. 7.5 Number of individuals with fitness equivalent to the best individual, up down oscillating series (top left), average best fitness (average monthly percentage return), up down oscillating series (top right), and the experiments price series (bottom)

with fitnesses equivalent to the best. This convergence of best individuals then goes on a steady increase after generation 30, which corresponds to a near optimal fitness.

7.6.5 Up down Oscillating Series

This series presents a compounded version of the series presented in the previous section, where the data set is made up of a series of up and down trends. In this case, a solution will ideally recognise the pattern and buy at each trough and sell at each peak. However, because the system adopts variable-size positions with a maximum possible daily investable amount, the system will have to begin buying slowly with smaller sizes as the series trends down with the biggest size executed at the bottom of the trough. It then scales back in size as the trend climbs.

In comparing the fitness and best individual convergence graphs with the previous section, similar curves are seen though in a smaller range. By the final generation for this price series, on average 7% of the population has converged to the best solution. For the simpler Up Down series this figure is 70% and 88% for the even more simple Up trending only series. An example of a best performer produced by GE for the final generation is provided below.

```
9100-MA( 14 )
```

Here a type of MACD oscillator is evolved. When the value of the moving average for the previous 14 days is greater than an index value of 9100 a sell signal is generated. The index increases and decreases in value by 100 points each day, meaning that a sell signal is generated 7 days after it was at a value of 9200. This causes the system to deliver a sell on the day the index value reaches 10000, presenting an optimal trade price. In building a position, the variable-size position taking plays a key role in the system achieving a greater average price. Here, the system begins buying with small sizes after the transaction costs amount to less than 20% when the index is at 7400, building up order size as the price decreases.

However, it should be noted that this rule, while producing an optimal return, is in fact very brittle. Over 100 generations, the population was only exposed to this one data set and as a result evolved rules that overfit specifically to this series. In much the same way as Lo [125] refers to humans developing heuristics based on the reinforcement and experience they have received through interactions with the market, so too does an artificial population of trading rules. If this rule were to be transferred to a dynamic environment, it would prove to be very inefficient and quickly accumulate losses as it encodes a specific index value in the expression. Given the context in which it was evolved such a rule is optimal, however, placed in a different or dynamic environment, and its actions will seem quite irrational.

Table 7.3 provides statistics on the trading performance of the system more generally. From this table it can be seen that, by the final generation, 94% of trades are profitable, on average returning 1004% return on the initial $10,000 cash position. While this figure may seem large in reviewing the price series, it can be seen that a position entered into at the bottom of a trough will return 100% if sold at the peak, which explains such large returns.

Table 7.3 Trading statistics for the Up Down Oscillating series

Metric	Score
Series(% return)	-45
Best(% return)	1280
Avg. Best(% return)	1004
Avg. No. Trades	267
Avg. Profitable Trades(%)	94
Avg. Losing Trades(%)	6
Avg. Fees($)	2730

7.7 Conclusion

This chapter has presented financial markets as a dynamic environment where the fitness landscape is variable and observes types of change that are complex in behaviour over time. Satisfying the criteria laid out by De Jong, and Trojanowski and Michalewicz, the financial markets represent an excellent test-bed for the testing of EC paradigms for dynamic environments.

A background to the pervading theories on market behaviour was provided. The EMH, Behavioural Finance, and the AMH were examined with the conclusion that inefficiencies can exist due to irrational or suboptimal actions taken by investors. Technical analysis was identified as a method for uncovering such inefficiencies and highlighting regularities or irregularities.

The trading model adopted in this book was described and underwent an initial testing phase on static data sets. Each of the goals of these experiments were satisfied, where it was shown that:

i. GE is capable of evolving sensible rules given the input data over a number of generations.
ii. For each controlled price series, GE was able to produce optimal or close to optimal rules.
iii. GE successfully produced rules that identified trends and their reversals using moving average indicators.
iv. These rules were accurate for the Up Down series and the Oscillating Up Down series as rules were produced where positions were entered into and closed at ideal prices.

Added to this, weaknesses of the static approach were observed where rules were produced that over-fit their specific data sets. Such rules, if taken out of context, would not yield the returns seen in training. To avoid such over-fitting, the population must not be trained over a single homogeneous data set but experience the full behaviour of a progressing price series.

The success of GE in these controlled static experiments now provides scope and a solid foundation for its application to a dynamic environment where the underlying data will change as time progresses with the expectation that more robust rules will be evolved.

Chapter 8
Adaptive Dynamic Trading with GE

The previous chapter conducted controlled trading over static artificial benchmark data sets. In these experiments GE was shown to be capable of producing optimal rules for trading. The problems of evolving over a static data set were highlighted as the resulting rules tended to be brittle and data sensitive. In this chapter GE is taken further and embedded in a dynamic moving window paradigm that evolves and adapts its population of rules over time.

The goal of this chapter is to examine the performance of GE when it is placed in a dynamic environment undergoing a Complex type of change, and to investigate whether it is capable of evolving robust, generalised, solutions. Chapter 4 identified BNF grammars and GE's many-to-one mapping as being key strengths in navigating dynamic environments. Chapter 5 demonstrated the benefits of BNF grammars where a more evolvable constant generation method, Digit Concatenation, was observed to adapt faster and more accurately. Chapter 6 took the exploration of BNF grammars further with the adoption of a meta-grammar and $(GE)^2$ that yielded benefits over standard GE in the dynamic problems examined. Here the benefits of GE's many-to-one mapping are highlighted where it is observed that a high level of diversity is maintained despite a large degree of evolution. It is the many-to-one mapping that is credited with providing GE with a good level of evolvability enabling it to adapt the population to new solutions where the environment changes significantly.

In order to demonstrate this an adaptive population of trading rules is evolved over price series in a moving window paradigm. A series of experiments are then undertaken culminating in an adaptive population being compared to a statically trained population on real-world price series. The experiments conducted are as follows:

- Initially an adaptive population is compared to a randomly generated population of trading rules on price series that have been shuffled clean of any technical trend information. This experiment highlights how an adaptive population of trading rules behaves when little or no information is available from the price series.

I. Dempsey et al.: Foundations in Gramm. Evol. for Dyn. Environ., SCI 194, pp. 141–161.
springerlink.com

- Next, the two setups are applied to real-world price series in order to see whether the adaptive population can outperform random search.
- Finally, the adaptive approach is compared to a static setup where a new population is generated and trained on the most up to date data exclusively before trading. These experiments serve to highlight the efficiency and utility of an adaptive approach over static analysis where the underlying data is dynamic in nature.

Added to this, evidence is presented that shows that GE is capable of uncovering information in historical price series to enable it to outperform a benchmark buy-and-hold strategy in a number of runs.

The next section explains the adaptive approach adopted in this chapter and Section 8.2 describes the diversity measurement used. Three sections of experiments follow where in the first two the adaptive approach is compared to a random trading setup and tested on a randomly shuffled data series in Section 8.3 and real-world price series in Section 8.4 on both the S&P 500 and the Nikkei 225. Section 8.5 compares the performance of the adaptive approach to that of a statically trained setup where the population is only trained over the latest window of data. Conclusions are presented in Section 8.6.

8.1 Adaptive Methodology

The adaptive GE methodology adopted firstly behaves as in earlier static studies that used the GE paradigm [22, 17, 152, 150, 149]. An initial training period is set aside on which the population of proto-trading rules is trained, with the aim that a competent population is evolved after a certain number of generations, G. The system then goes 'live', and a number of actions are taken by the system. The trading system takes the best performing rule from the initial training period, and uses this rule to trade for each of the following n days. After n days have elapsed, the *training window* moves forward in the time series by n days, and the population is retrained over the new data window for a number of generations g, where $g < G$. This embeds both a *memory* and an *adaptive potential* in the trading system, as knowledge of good past trading rules is not completely lost, rather it serves as a starting point for their subsequent adaptation.

A small value of g means that memory is emphasised over adaptation, as the new data has relatively less chance to influence the trading rules. This could be considered a tuning parameter that could be used to alter the adaptive characteristics of the system, and could itself be open to adaptation. For example, in periods of rapid market change a trading system with a 'long memory' could be disadvantageous, whereas in stable periods, a longer memory could well be advantageous. Similar comments can be made for the parameter n. If its value is large, the trading rules are altered less frequently, but each adaptive 'step' will tend to be larger.

In implementing the moving window training process in this case, the first 250 days data is used to create the initial population of trading rules. Seventy five days of data are reserved at the beginning of the time series as an initial buffer to allow the evolved rules use moving averages of up to a maximum lag of 75 days for the first day, extending as time progresses. The trading rules are trained on the data that follows, for 100 (G) generations. The trading rule that generates the best return over the training period then trades 'live' (out of sample) for the next 5 days (x). Following this the training window is moved forward to include these 5 days, and the population of trading rules is adapted by retraining it for 10 (g) generations over the new window. Figure 8.1 provides a diagram of the training/live trading process.

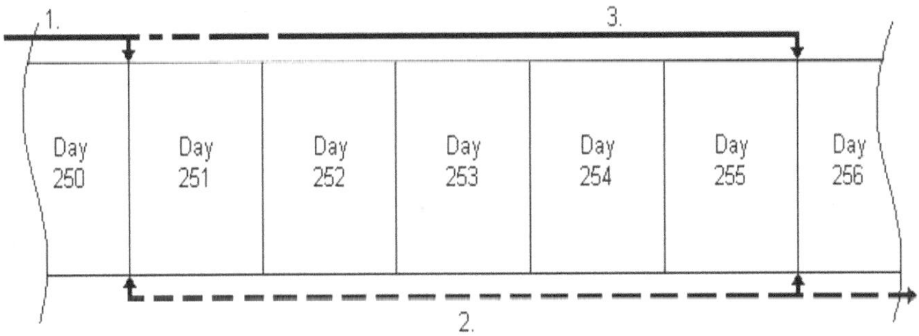

Fig. 8.1 The initial population of trading rules is evolved using data from days 1-250 (step 1 in the diagram). Next the best of these rules is used to trade live for five days (step 2). Finally, the training window is moved forward five days in the dataset, and the current population of rules is retrained for g generations (step 3).

8.2 Diversity Measurements

In Section 3.3, the maintenance of a diverse dispersed population was identified as playing a key role in helping populations adapt to changes and discover new solutions in areas of the solution space that had previously yielded poor fitness. In GE, a genotype-to-phenotype mapping takes place that leads to a many-to-one mapping from the genotypic space to phenotypes. As such, two levels of diversity are observable:

i. **At the genotypic level:** This can be analysed by examining the number of distinct genetic makeups within the population.
ii. **At the phenotypic level:** This can be analysed by examining the number of distinct fitnesses within the population.

This chapter focuses on the latter as evidence suggests [188, 101, 36] that genotypic diversity measurements do not accurately capture population dynamics and that phenotypic diversity metrics correlate better with run-time performance. The nature of GE's representation also points towards phenotypic measurements as a many-to-one mapping combined with variable length genotypes make convergence on a genotypic level difficult to achieve.

Considering this the number of distinct phenotypes is logged, where a distinct phenotype is a unique fitness. Combined with this the entropy of the population is measured. Rosca [185] used Shannon's information entropy formula to encapsulate Koza's observation that histograms represent an informative view of the progress of the learning process for a population. Entropy is calculated by:

$$E = -\sum_k p_k.log(p_k) \tag{8.1}$$

where p_k is the proportion of the population P occupied by population partition k. In this case k is a distinct fitness and so, the population is divided up into partitions in this manner.

8.3 Shuffled Data Sets

This round of experiments was conducted using shuffled data sets where two real-world time series, the S&P 500 and the Nikkei 225, were randomly shuffled. The adaptive system was then applied to this data set using the same parameters as the previous chapter along with the random setup. The purpose of this is to explore what type of behaviour is evolved, if any, where an adaptive population is evolved over series that are devoid of any technical trend information. The absence of any technical trend information in a series removes the key driver to the Moving Average indicator that is designed to uncover trends. The results of the adaptive population are then compared against a random system. The random trading rules are selected through creating a new population using the same mechanisms as the first generation of a normal run. The first mappable rule from this population is then selected to trade. In doing this the adaptive population is compared to random trading. If performance and behaviour correlates between the two approaches then the adaptive population can essentially be said to be conducting a random search and achieving any returns through "luck". On the other hand if the performance and behaviour diverge, then GE is seen to learn from the underlying data.

A population size of 500 individuals was used with 100 generations of training for the initial period. Similar to experiments conducted in Chapter 7 and previously, a generational rank replacement strategy was used with 25% of the weakest performing members of the population being replaced with newly-generated individuals in each generation. Thirty runs were conducted for each of the experiments for each market. Table 8.1 provides a summary of the system parameters.

Table 8.1 Experimental parameters adopted

Parameter	Setting
Population Size	500
Crossover	0.9
Mutation	0.1
Selection	Roulette
Replacement	25%
Wrapping	8 events
Codon Size	8 bits
Bank	$10,000
Interest	US Treasury & Japanese Treasury
Max Order Size	$1,000

8.3.1 S&P 500 Shuffled Data Set

A program was employed to shuffle the S&P 500 data series with the aim of essentially removing any technical trend information that could be obtained from evaluating historical data. The adaptive and random systems were then applied to this data set using the same parameters as in Table 8.1. Table 8.2 displays a break down of the trading activity between the two set ups.

Table 8.2 Trading statistics for adaptive and random systems on the shuffled S&P 500

	Adapt	Random
Best(% return)	372	1773
Avg. Best(% return)	19	611
Avg. No. Trades	147	738
Avg. Profitable Trades(%)	48	49
Avg. Losing Trades(%)	52	51
Avg. Fees($)	1648	8258

8.3.2 Discussion

In examining these results it can be seen that the random approach outperforms the adaptive approach significantly. The adaptive approach is also seen to execute significantly fewer trades. This difference is likely driven by the fact that trading costs are the only consistent feedback the system has from trading. The random approach which has no knowledge of trading costs accumulates five times the fees. A factor which played to the random system's advantage is that for the shuffled S&P 500 data set the majority of data points are located in the bottom third of the range which when combined with the fact that the system invests a fixed amount as opposed to buying

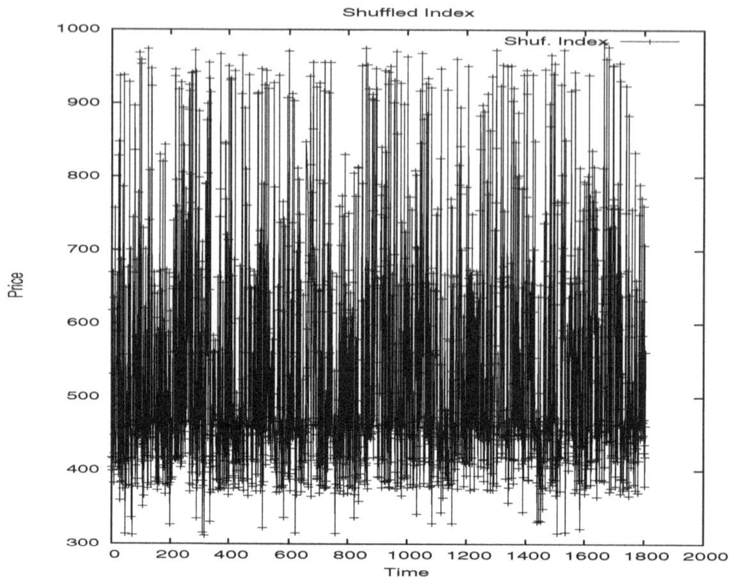

Fig. 8.2 Shuffled S&P 500 series

a number of shares at a certain price would mean that the down side is less than an upside over the same range. If \$1000 is invested at 20,000 and the price falls to 10,000 this is a loss of 50% but \$1000 invested at 10,000 which goes up to 20,000 yields a return of 100%. This then rewards frequent trading where most trades wont generate a loss as most of the data points are in the bottom tier but the odd lucky trades that buy in when the price is on the average low are highly rewarded when sold on a high.

An argument that then arises is: why did the adaptive approach not learn to trade as frequently as the random setup if the rewards are higher? Considering how the moving average technical indicator works, it is actually difficult for the system to evolve a rule that would generate a high frequency of trades. This is because both the long and short moving averages in the shuffled data set would be equivalent at any point in the series, highlighting a weakness in the grammar provided. Whereas the random approach will generate buy and sell signals randomly resulting in a high frequency of trading. Figure 8.2 displays the shuffled S&P series.

8.3.3 Nikkei 225 Shuffled Data Set

Again the same parameters were used here as in previous sections and both systems were executed on the shuffled Nikkei 225 data series. Table 8.3 displays the comparison.

In this case the comparison between the two systems is more even. This is primarily down to the higher density of data points being in the middle as

Table 8.3 Trading statistics for adaptive and restart systems on the shuffled Nikkei 225

	Adapt	Random
Best(% return)	36	257
Avg. Best(% return)	17	1
Avg. No. Trades	60	292
Avg. Profitable Trades(%)	57	51
Avg. Losing Trades(%)	43	39
Avg. Fees($)	674	3351

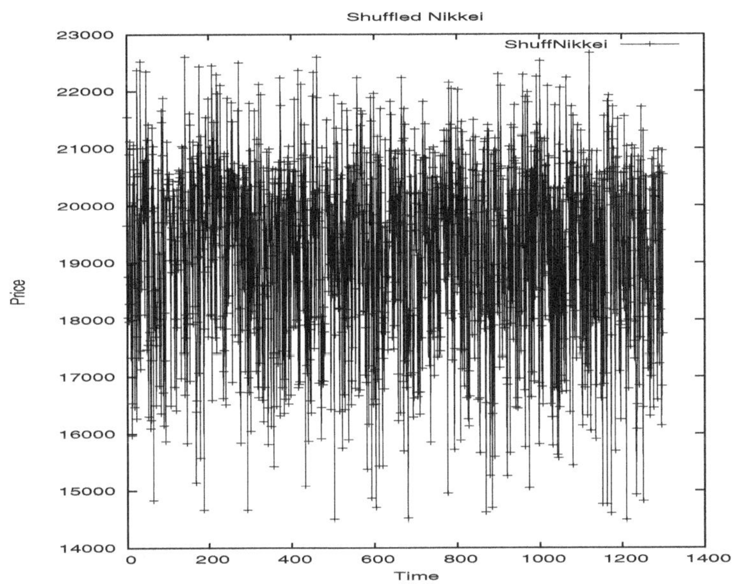

Fig. 8.3 Shuffled Nikkei 225 series

seen in Figure 8.3. Again the random system conducts much more frequent trading, coming in at almost five times the number of trades of the adaptive system.

8.3.4 Discussion

From these results it can be said that the performances of both setups are fairly even. The differences in return are discounted by the larger trading costs incurred by the random approach that do not feed back into the systems learning. In the shuffled Nikkei data set the points are distributed more normally with less of a bias towards the bottom third as seen in the shuffled

S&P 500. Consistent behaviour across both data sets is seen in the number of trades executed where in both cases the random setup executes five times the number of trades that the adaptive approach performed.

This suggests that the adaptive approach is learning from the only reliable signal it is receiving from training on these series, that trading carries cost and risk. With this information the adaptive population adopts a more conservative trading strategy, choosing to trade less. While such a strategy did not outperform random trading on the shuffled S&P 500, for the shuffled Nikkei 225 where the data is distributed more normally the more prudent strategy yields better results.

8.4 Adaptive vs Random

In the previous experiments an adaptive population was seen to evolve a conservative trading strategy where there was an absence of any technical trend information in the time series. Here the adaptive population is again compared against the random trading population but now it is applied to real-world price series. The purpose of these experiments is to confirm the hypothesis that GE is indeed extracting useful information from a dynamic time series. To achieve this, these experiments focus on comparing the 10 generational adaptive population against random trading rules. Again a divergence in behaviour and performance is sought between the adaptive and random trading setups. In these experiments the real un-doctored price series is used and so consistent superior performance on the part of the adaptive population can be attributed to the system actually mining information from the price series and learning to trade better based on technical information.

The setups were applied to two indices, the S&P 500 for the period January 1st 1991 to December 1st 1997, and the Nikkei 225 for the period December 10th 1992 to December 3rd 1997 (Figures 8.4 and 8.5 display the S&P 500 and the Nikkei 225 over the relevant time periods). All experimental parameters for the adaptive population are the same as that in the previous section of experiments and Table 8.1. Experiments are conducted on the real historical S&P 500 and the Nikkei 225. It should be noted that while one cannot trade these indices directly, Equity Traded Funds (ETFs), futures, and over the counter contracts can be traded that reflect the movements and returns of the indices.

8.4.1 S&P 500 Results

Table 8.4 outlines the comparison of performances between the random and adaptive approaches for the S&P 500. For these experiments the Sharpe Ratio is also provided. The Sharpe ratio [199] is a commonly used ratio for

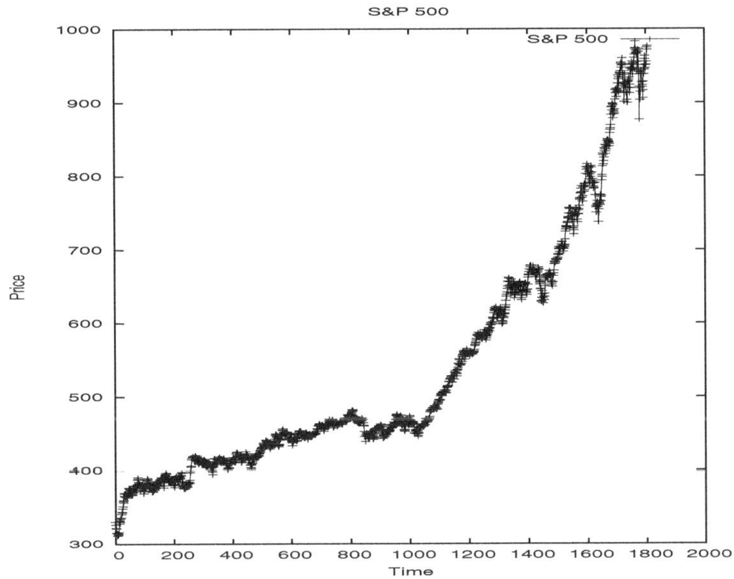

Fig. 8.4 The S&P 500 Index 1/1/1991 to 3/12/1997

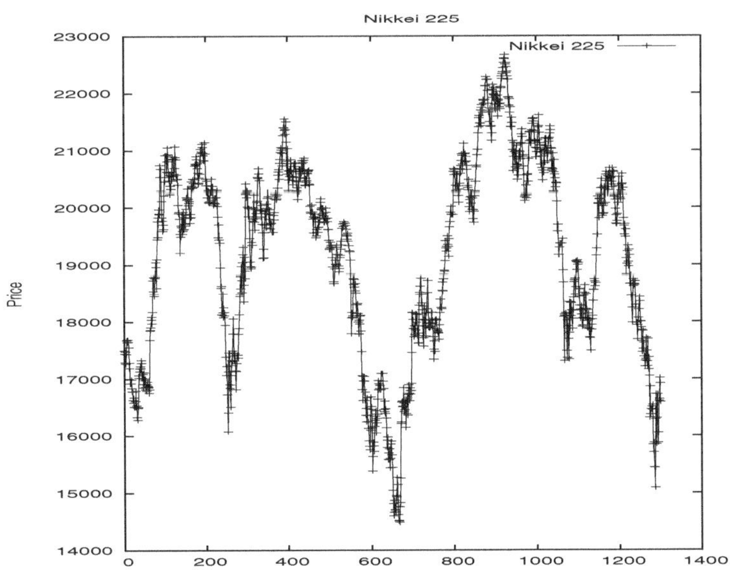

Fig. 8.5 The Nikkei 225 Index 10/12/1992 to 3/12/1997

evaluating the risk-adjusted returns of an investment where the excess returns of the investment are divided by the volatility of the returns. This was calculated as the monthly rate of return less interest divided by the standard

Table 8.4 Trading statistics for adaptive and random approaches on the S&P 500

	Adapt	Random
Index(% return)	134	134
Best(% return)	111	54
Avg. Best(% return)	107	25
Avg. No. Trades	73	432
Avg. Profitable Trades(%)	71	68
Avg. Losing Trades(%)	29	32
Avg. Fees($)	804	4801
Avg. Sharpe Ratio	18.64	-2.14

deviation of the returns. In this statistic it can be seen that the adaptive setup again out performs the random trading performance.

8.4.2 Discussion

In reviewing these results it can be seen that three statistics stand out in their contrast:

i. The adaptive population outperforms the random trading rules by a significant margin, providing on average four times the return.
ii. While the ratio of profitable to loss making trades under both setups is similar, the degree to which the adaptive population is more profitable is far greater. On each trade the adaptive approach made an average of $140 where as for the random approach it was an average of just $14, 10 times less.
iii. Because the random population receives no reinforcement from its actions, it has no notion of transaction costs or experience of loss through trading. Due to this the random approach is again seen to conduct a large amount of trades as seen through the fees it accumulates. The random approach on average experienced 6 times the amount of transaction costs compared to the learned conservative trading of the adaptive approach.

These points taken together suggest that the adaptive population is learning to trade more conservatively while at the same time choosing more optimal points to enter and close positions, evidenced by the far superior average return per trade. This carries the implication that GE is able to extract information from the time series so that it may trade more effectively.

In terms of performance against the benchmark buy-and-hold, while the adaptive approach did not outperform the benchmark it did provide excess returns of over 100% on average with just 4% between the best performer and the average best over all runs. Indeed outperforming the index in this case is a difficult task given the aggressive fee schedule and strongly up trending price index (Figure 8.4) leaving little room to maneuver in closing out of a long position to reinvest at a lower price.

8.4.3 Nikkei 225

Table 8.5 outlines the comparative results for the adaptive and random trading approaches on the Nikkei 225.

Table 8.5 Trading statistics for adaptive and random strategies on the Nikkei

	Adapt	Random
Index(% return)	-17	-17
Best(% return)	2	-12
Avg. Best(% return)	-21	-35
Avg. No. Trades	63	259
Avg. Profitable Trades(%)	25	46
Avg. Losing Trades(%)	75	54
Avg. Fees($)	701	2911
Avg. Sharpe Ratio	-8.18	-10.48

8.4.4 Discussion

Again in these experiments the adaptive system outperforms the random approach by a significant margin despite the random system, in this case, producing more profitable trades on average than the adaptive. However the extent of the losses incurred by the random setup suggests it engages in riskier trading again due to the fact that it receives no reinforcement from its trading. This is also borne out by an inferior Sharpe ratio compared to the adaptive setup.

Of particular importance in these experiments is the performance of the adaptive approach versus the benchmark buy-and-hold of the Nikkei 225 for the same period. As can be seen in Table 8.5, the best run from these experiments made a positive return of 2%, 19% greater than the benchmark. More than this however, the adaptive approach beat the benchmark in 23% of the runs conducted compared to just 3% for the random setup. Considering evidence observed in the previous chapter where GE demonstrated an ability to uncover trends in static price series, and in the previous section of this chapter where an adaptive population was demonstrated to learn to trade more conservatively in the absence of technical trend information, and given the divergence in performance from random trading it can indeed be said that the adaptive population is generating these returns through uncovering information learned from the time series. This then suggests that the Nikkei 225 is not entirely efficient.

These results have confirmed that the adaptive GE setup is indeed learning information through training across the data set. Training across the dynamic data set equips the adaptive approach with a more conservative trading style and also an ability to identify more optimal points at which to trade.

8.5 Adaptive vs Restart

This round of experiments examines the premise that adopting an adaptive approach, whereby a persistent population is continuously evolved over time across the data set, is indeed beneficial over a population that is only evolved over the current data exclusively. In order to do this the experiments compare the performance of the adaptive setup as seen in the previous section compared to a population that is evolved statically over the current moving window. The static setup destroys its population from the previous trading window and generates a new one when the window moves forward. This population is then trained from scratch for 10 generations on the current window, investing equal computational effort as compared to the adaptive approach. Thus the static setup has no memory or exposure to data outside the current window.

The rest of this section is broken up into two parts. Section 8.5.1 reports the returns of each trading system over its training range versus the returns made by the index over the same period. Section 8.5.2 reports the returns made during out-of-sample trading versus the return to the indices over the same period.

8.5.1 Training Returns

As the training window moves forward in the data set by 5 days on each increment, the S&P 500 dataset produced 294 distinct retraining windows, and the Nikkei 225 dataset produced 192 distinct training windows. At the end of the final generation in each training period, the return of the best trading rule (based on its fitness over that training window) was determined.

Figures 8.6 and 8.7 display the average monthly return over the preceding 260 days of training for the setups under investigation on both indices. As can be seen from these graphs, the in-sample returns of both setups are very similar with no sustained divergences in experiments on either index.

As a large degree of evolution occurs in the case of the adaptive population as it is retrained across the price series there is potential for convergence or significant drop in diversity. Figures 8.8 and 8.9 compare the entropy of the adaptive and static populations at the close of each training period. In order to understand the behaviour of the entropy series the relevant index is also charted on the same graph along the right hand y-axis. Three pieces of information can be drawn from these graphs:

i. It can be noted that the adaptive population does not progressively converge as the training window moves across the price series. By the end of the data sets in both cases, the entropy of the adaptive population is almost the same as it was at the end of the first training window of the experiment.

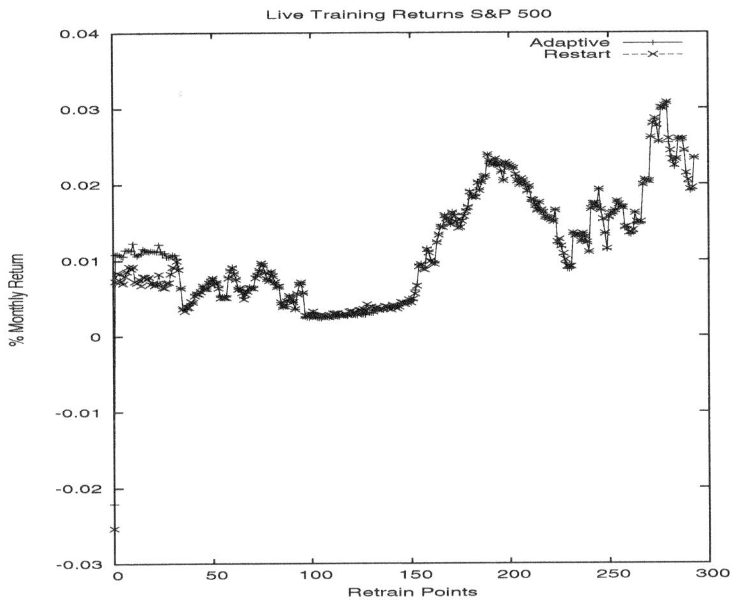

Fig. 8.6 Training performance on the S&P 500 for the adaptive and static populations

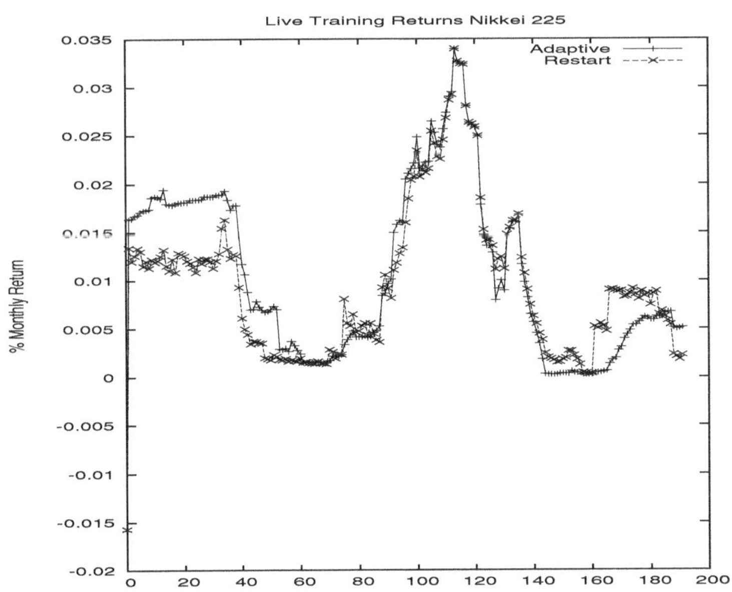

Fig. 8.7 Training performance on the Nikkei 225 for the adaptive and static populations

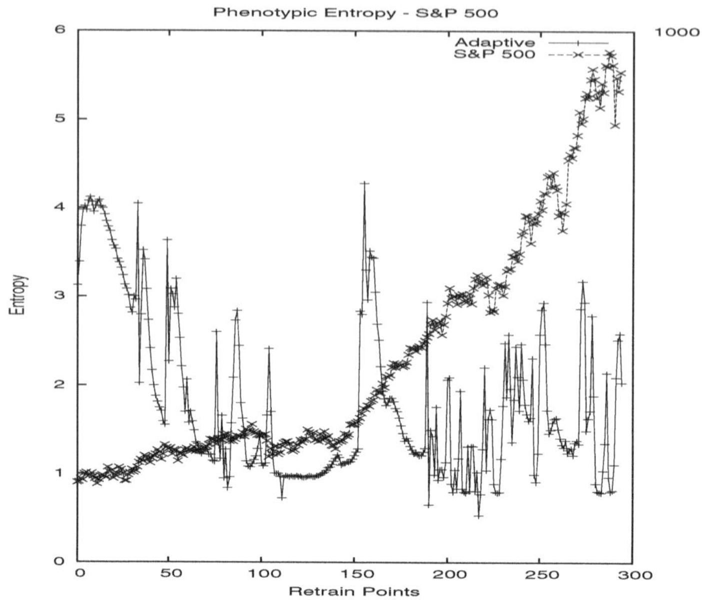

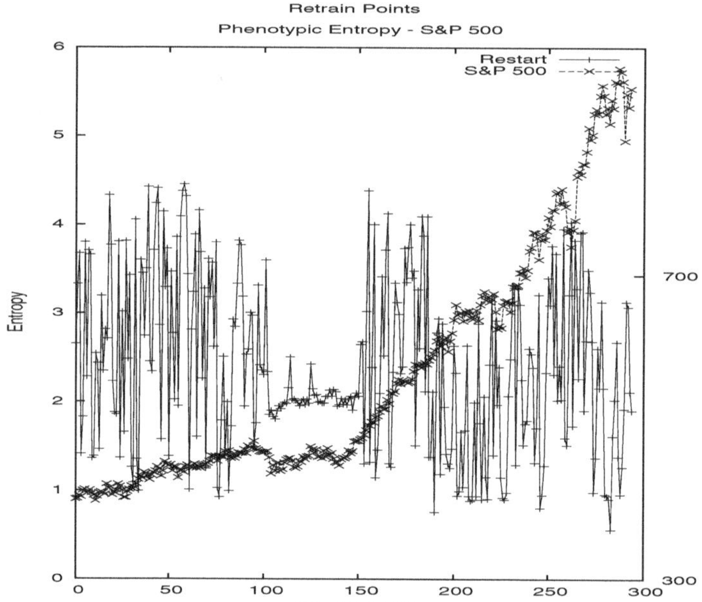

Fig. 8.8 The entropy of both the adaptive (top) and static (bottom) populations at the final generation of each training window with the entropy value along the left-hand y-axis. The S&P 500 index is also charted for the same dates with its values along the right-hand y-axis.

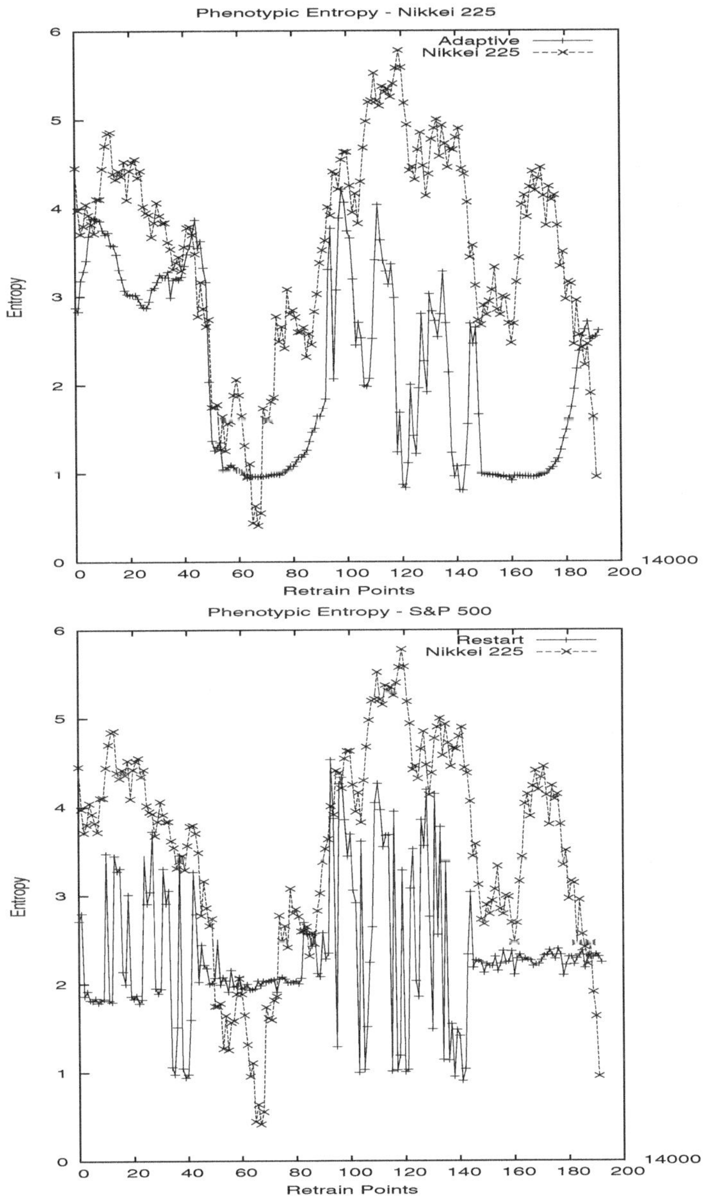

Fig. 8.9 The entropy of both the adaptive (top) and static (bottom) populations at the final generation of each training window with the entropy value along the left-hand y-axis. The Nikkei 225 index is also charted for the same dates with its values along the right-hand y-axis.

ii. Where the index is making a loss over the preceding year (training window) the entropy of both populations drops and levels out. In situations where the index is making a loss, as seen in the Down series of Section 7.6.2, it is in the interests of the system to pull out of the market and stay in cash. This drop in phenotypic entropy where the index makes a loss suggests that both setups respond correctly to the environment and swap to cash positions generating fitnesses that would have a low degree of diversity as they would simply be returning the gains received through interest.
iii. Outside of the areas where the indices are making losses entropy is seen to increase significantly and indeed become volatile as the populations evolve over landscapes that provide multiple different opportunities for trading profitably.

Key to these graphs is that diversity is seen to be a function of the problem landscape. When the index is making a loss only one solution presents itself – to maintain a cash position, as the system is not permitted to take short positions. However, when the index provides a positive return or is volatile, entropy within both setups is seen to increase. This is of significant importance to the adaptive population as, from the beginning to the end of the data sets, this one population undergoes 2,940 generations of evolution for the S&P 500 and 1,930 for the Nikkei 225. The ability to maintain this level of diversity throughout a run is down to the many-to-one mapping in GE from genotype to phenotype. This allows the evolution of neutral networks and provides a high level of evolvability, enabling the evolution of new solutions quickly.

For informative purposes the number of distinct genotypes was also logged and this count remained at the population size across all runs – always 500 distinct genotypes. This represents a very high degree of diversity in the genotypic space. Considering diversity converges and fluctuates in the phenotypic space this suggests that neutral networks are being explored among the genotypes, continuing the evolutionary search while maintaining stable phenotypes.

8.5.2 Out-of-Sample Returns

Table 8.6 reports the results for the adaptive versus static approach for the S&P 500.

The adaptive approach is seen to produce far superior results over the static setup producing, on average, over 3 times the return. Similar statistics are reported for the static setup compared to the random setup of previous sections with much more trading being conducted with very high transaction costs being a result, in the region of five times the adaptive approach's total average fees.

In Table 8.7 the trading statistics for the Nikkei 225 are reported. In the previous section the adaptive setup was seen to outperform the benchmark

Table 8.6 Trading statistics for adaptive and static approaches on the S&P 500

	Adapt	**Static**
Index(% return)	134	134
Best(% return)	111	58
Avg. Best(% return)	107	27
Avg. No. Trades	73	379
Avg. Profitable Trades(%)	71	68
Avg. Losing Trades(%)	29	32
Avg. Fees($)	804	4199
Avg. Sharpe Ratio	18.64	0.011

Table 8.7 Trading statistics for adaptive and static strategies on the Nikkei

	Adapt	**Static**
Index(%)	-17	-17
Best(%)	2	-11
Avg. Best%	-21	-29
Avg. No. Trades	63	232
Avg. Profitable Trades(%)	25	47
Avg. Losing Trades(%)	75	53
Avg. Fees($)	701	2589
Avg. Sharpe Ratio	-8.18	-7.78

buy-and-hold in 23% of the runs conducted. For the static approach 14% of runs outperform the benchmark. Despite this the average performance of the static approach is again inferior to that of the adaptive though not to the same scale as on the S&P 500. In this case the adaptive approach is seen to produce a slightly inferior Sharpe Ratio.

8.5.3 Discussion

In viewing these results together it can be seen that in training both setups produced similar returns yet their results diverged when trading progressed to out-of-sample data. This suggests that the adaptive setup is indeed producing more robust generalised solutions. To demonstrate this Table 8.8 displays the percentage of times that a best performing rule survived as best performer across more than one interval.

Table 8.8 Percentage of rules that survive more than one increment

S&P 500	**Nikkei 225**
49%	52%

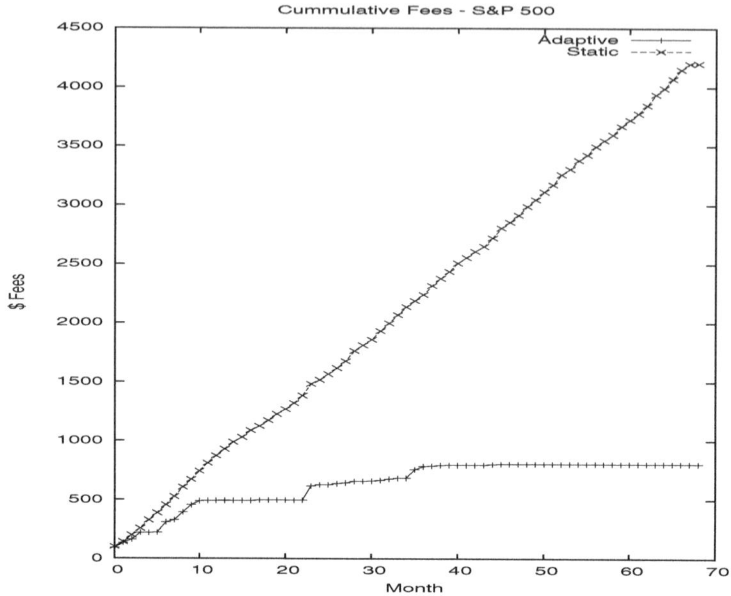

Fig. 8.10 Accumulative fees incurred by adaptive and static setups on the S&P 500

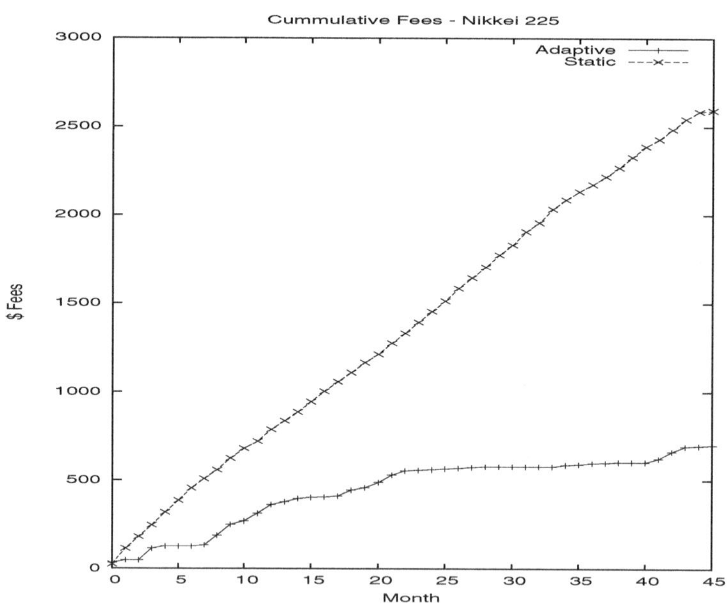

Fig. 8.11 Accumulative fees incurred by adaptive and static setups on the Nikkei 225

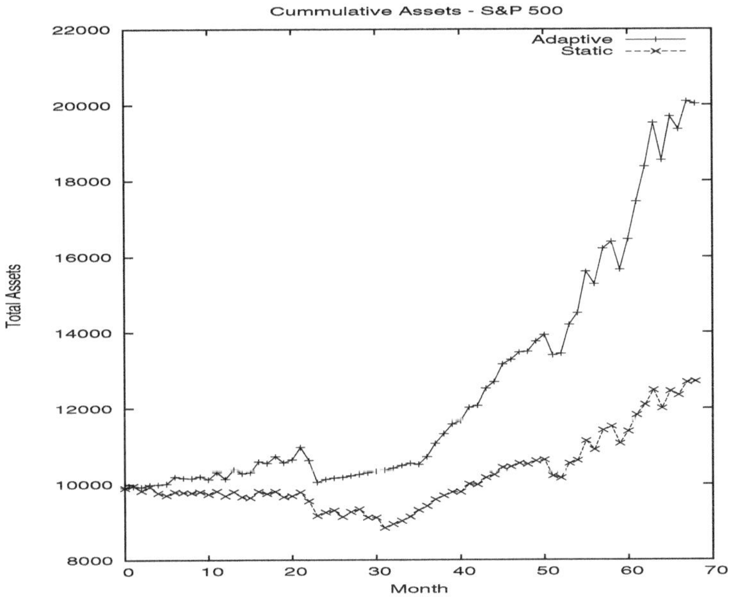

Fig. 8.12 The total average assets being managed by the adaptive and static setups for the S&P 500

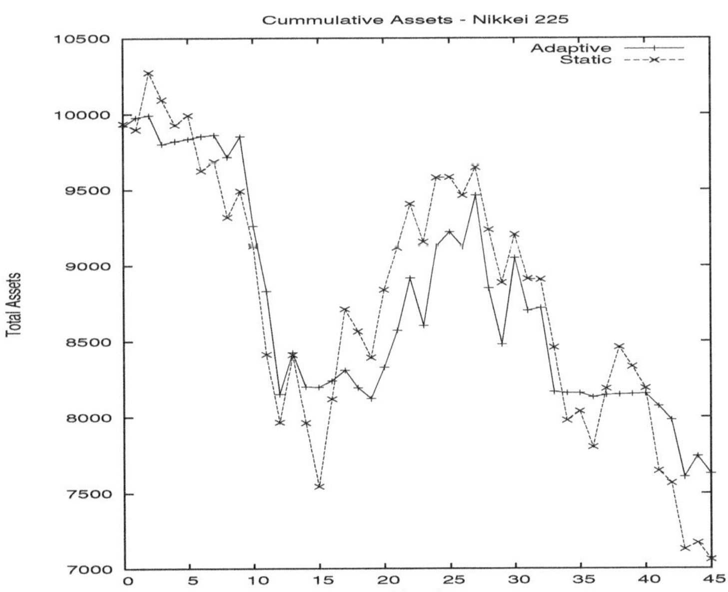

Fig. 8.13 The total average assets being managed by the adaptive and static setups for the Nikkei 225

This highlights a major strength in the adaptive approach, that it is capable of producing generalised rules as it is exposed to more diverse data to the extent that on average a rule will bridge two increments. In the case of the static approach a new population is generated at each increment. Considering that the in-sample return of both approaches was to a large extent similar it may be an indication that the static approach is in fact over-fitting to the training data, a point that would tie in with experiments conducted in the previous chapter where brittle rules were seen to be produced on static data.

The more generalised perspective of the adaptive approach is also the reason behind it incurring far less transaction costs than the static approach. To appreciate this better Figures 8.10 and 8.11 display the accumulative fees incurred by both setups on each index.

It can be seen that for the S&P 500 the adaptive approach essentially evolves a buy-and-hold strategy, incurring little in the way of fees after the 35th month, while the static approach continues to trade at an almost equal rate through out. For the Nikkei 225 the adaptive approach again trades much more conservatively this time conducting trading right across the time series as it seeks to take advantage of the dip and recovery of the Nikkei 225 data set towards the end of the series. These graphs demonstrate a more complete understanding of the underlying data series on the part of the adaptive setup when compared to the static approach which interacts with the market at the same rate through out the series.

These trends are also highlighted in the total average monthly position being managed by both setups in Figures 8.12 and 8.13. This figure includes cash and the current value of investments in the index.

8.6 Conclusions

This chapter progressively investigated the behaviour of the adaptive population over three sets of experiments. Section 8.3 saw the adaptive approach compared to random trading on data sets that had been shuffled clean of any technical information. From this section the adaptive approach was seen to adopt a more conservative trading strategy and trade a lot less than the random approach indicating that it was able to learn from the dynamic data through consistent feedback on trading costs despite the underlying data being devoid of any trend information. Essentially this section demonstrated that despite the price series being largely random the adaptive population was able to extract information and do what was in its best interest – trade less.

Section 8.4 progressed the adaptive population on to the real-world time series where it was again compared against a random trading strategy. This section saw the adaptive population demonstrate an ability to extract meaningful information from the price series. On these data sets it identified more points at which to trade on the S&P 500 generating far superior returns and on the Nikkei 225 was able to outperform the benchmark buy-and-hold in 23% of the runs conducted. While this is a positive result it should be noted that the buy-and-hold benchmark does embark on a downward turn as the Nikkei 225 data set comes to a close. Continuing from the previous experiments the adaptive population was again seen to execute far fewer trades than the random setup which in turn led to lower transaction costs resulting in more funds for investment and a better overall return.

The adaptive approach was compared to the performance of a statically trained population in Section 8.5. This section conducted an analysis of diversity at the phenotypic level of both setups with the entropy of each measured at the final generation of each training window. The key message here is that the entropy or phenotypic diversity of both setups was a function of the current state of the environment. Despite a large level of evolution occurring for the adaptive population it did not converge but fluctuated in response to changes in the environment. When the universe of potentially fit solutions converged as the indices made losses, so too did the entropy of the populations. However as opportunities presented themselves to generate profits, outside of cash positions, entropy was seen to increase. The ability for GE to achieve this represents a major strength in its application to dynamic environments and is a result of the high level of evolvability presented through the genotype-to-phenotype mapping.

In comparing the out-of-sample results of the adaptive and static setups, the adaptive approach was seen to consistently outperform the static. This is despite both setups demonstrating similar returns in training. The previous chapter provided evidence of a statically trained population producing brittle results which overfit the given training data. In these experiments this again is the case for the statically trained population. On the other hand the adaptive setup was able to produce robust generalised solutions that were evolved over a heterogeneous data set, as the training window and population progressed across the time series. Indeed, robust solutions were seen to be evolved in 50% of window increments. This more macro view of the data drove the adaptive population to invest at more appropriate times as evidenced by the cumulative fees graphs. In the case of the S&P 500 the adaptive approach evolved a buy-and-hold strategy as the index embarked on a sustained bull run. For the Nikkei 225 fees were generated at points where the market presented opportunities for investment. In contrast the static approach essentially traded in and out of the market at a sustained pace right across the time series without demonstrating the more macro perspective of the adaptive population.

This chapter also served to highlight the utility of using a real-world problem as a testbed for dynamic approaches to EC. The financial domain presents an environment that continuously undergoes complex change with a wealth of standardised data readily available. The presence of domain specific analytics and metrics of performance such as the Sharpe Ratio, also provide the modeller with further standardised insights into the behaviour of the system under investigation.

Chapter 9
Conclusions and The Future

9.1 Summary

This book began by posing three questions concerning the application of GE to dynamic environments.

i. Is the correct infrastructure in place for GE to navigate dynamic environments? This required the investigation of the potential strengths inherent in GE and areas that require further attention for the effective application of GE to these environments
ii. Is GE capable of discovering new solutions when change occurs in the environment? Fundamental to the navigation of dynamic environments is that a population of solutions be capable of quickly transitioning to new areas of the solution landscape.
iii. Can GE maintain a diverse population of robust solutions capable of handling dynamic data? A major criticism of GA/GP approaches for dynamic environments is their tendency to converge. Maintaining diverse robust solutions is imperative for the successful application of GE to dynamic environments.

Combined with these, questions were posed relating to wider EC issues on the roles of memory and diversity in dynamic environments, along with how EC paradigms might best be tested in order to investigate these questions.

Following an introduction to Grammatical Evolution, a comprehensive survey of the state of the art for EC in dynamic environments was conducted. Stemming from this survey, a clear picture of the types of dynamic environments possible was outlined, along with the unifying of a number of researchers' work that identified the types of change that can occur in such environments. The various approaches researchers have adopted to navigate dynamic environments and overcome the extra issues specific to dynamic problems, were identified. Included in this was a discussion of performance measurement in the context of dynamic environments. An analysis of prior research was developed, producing a list of open research gaps.

I. Dempsey et al.: Foundations in Gramm. Evol. for Dyn. Environ., SCI 194, pp. 163–169.
springerlink.com

We then set out by providing an examination of GE in the context of dynamic environments, and reported our findings in some early experiments adopting non-stationary symbolic regression instances. Out of this, key strengths and areas for improvement were identified. Critical to the successful evolution of solutions for dynamic environments is an efficient mechanism for the generation and adaptation of constants, a mechanism that was lacking in the current state of the art in GE. A number of potential strengths were also identified, along with benefits unique to GE brought about through its use of BNF grammars.

Two novel constant generations mechanisms were introduced for GE or GP-type paradigms. The study then embarked on a series of experiments aimed at investigating the different methods of constant generation and adaptation. These experiments sought to focus on the specific issue of constant creation and adaptation in isolation. Both static and dynamic problems were examined with three different types of constant generation methods, along with experiments where the evolutionary search itself was used to determine the best method. This lead to the identification of Digit Concatenation with an ability to form expressions as the most efficient mechanism. Digit Concatenation presented an evolvable representation at the grammar level that provided advantages such as greater accuracy and incremental evolution, over the other methods explored.

Further exploration of the benefits of BNF grammars was then conducted with the introduction of another novel method of constant generations through the use of meta-Grammars and $(GE)^2$. Experiments were executed with the $(GE)^2$ paradigm using a population half the size of that used in prior experiments, so as to maintain an equal computational effort. Despite the reduced population size and an increase of the search space, brought about through the solution-grammar chromosome, $(GE)^2$ performed competitively with standard GE, while also demonstrating an increased rate of evolution. For dynamic experiments, $(GE)^2$ was seen to adapt at a greater rate over some prior experiments showing significant potential for further exploration.

Having completed this, experiments were then executed aimed at gaining a wider understanding of the behaviour of GE in the complex setting of a real-world problem, that of trading on financial markets. A background to modern financial theory was first presented to provide a backdrop to the significance and complexity of the domain. This was followed with a survey of prior work carried out in applying EC techniques to the domain. Stemming from this, a number of controlled artificial static experiments were conducted. This approach was adopted in order to gain a clearer understanding of the system's behaviour under a noise-free setup. These experiments demonstrated the ability of GE to perform in the domain and also highlighted weaknesses of the static approach.

This then provided the scope to apply GE to dynamic financial data sets. A variety of experiments were conducted that sought to build up complexity, allowing a greater understanding of GE's behaviour to be developed at each level. First GE was seen to trade more conservatively than random trading for shuffled data sets. Progressing onto the normal unshuffled data, the adaptive approach yielded superior returns over random trading. Finally in comparison to a static approach, the adaptive approach also produced superior results. In addition these experiments also served to highlight GE's ability to maintain diversity within the population as a function of the potential fitnesses in the search space.

Arising from this book, these series of experiments provide evidence for the benefits of novel constant creation methods. It identifies evolvability and genotype-to-phenotype mapping as being key assets for evolutionary algorithms applied to dynamic environments, circumventing issues surrounding the fundamental problem in the literature thus far, that of convergence and reduction in diversity.

9.2 Research Results

Through the gathering together of the various literature on the topic of EC in dynamic environments and the subsequent execution of a series of experiments in the domain, a number of key insights are provided on different levels. This section will review the research results of this book.

9.2.1 Analysis of GE in Dynamic Environments

This book for the first time analysed GE in a dynamic environment. The issues surrounding evolution in dynamic environments are different to those of a static one. In Chapter 4, key potential strengths of GE for dynamic environments, along with areas where it could be improved were identified and analysed alongside work in other areas of EC. Subsequent chapters then introduced effective evolvable methods for constant generation and adaptation tailored for dynamic problems; examined the diversity-maintaining effects of the genotype-to-phenotype mapping; analysed $(GE)^2$ in dynamic environments; and demonstrated the effectiveness of GE in dynamic environments. Combined with this, as far as the authors are aware, this book represents the first comprehensive investigation of a GP-type paradigm set in a dynamic environment.

9.2.2 Comprehensive Literature Review for EC in Dynamic Environments

Chapter 3 provides a detailed survey of the various facets and issues surrounding the application of EC techniques to dynamic problems. The chapter began by questioning what constituted a dynamic environment and what the possible types of change were that could occur. This led to the unifying of a number of author's works to form a new taxonomy that then provided a foundation for analysis of the various approaches researchers have adopted in extending EC to handle dynamic environments. Comprehensive coverage was given of the various approaches taken by researchers. This was followed by an examination of the different metrics adopted to measure performance over dynamic landscapes as well as the types of benchmarks used in evaluating approaches. This chapter uncovered a number of gaps in the state of the art, where very little attention is paid to GP-type paradigms in dynamic environments, and too much attention is paid to simplistic benchmarks and equipping algorithms with forms of explicit memory.

9.2.3 Extensions to GE's Ability to Create and Adapt Constants

The majority of phenotypic solutions generated by evolutionary algorithms contain constants. Where the environment is dynamic, there is then a requirement to generate new and/or adapt existing constants. Thus, an effective and efficient mechanism for creating and adapting constants is imperative. GE and GP generally adopt an inflexible approach to the creation and adaptation of constants. This book explored the properties and flexibility of BNF grammars, unique to GE, to create efficient and flexible constant generation and adaptation mechanisms. Issues such as ease of evolution and accuracy were uncovered and addressed. Digit Concatenation with the ability to form expressions was identified as being the best method for constant creation and adaptation.

9.2.4 Novel Methods for Constant Creation in GE

In the process of investigating and identifying the best method for constant creation, two novel methods for constant creation were introduced: Digit Concatenation, Persistant Random constants, and constant generation through $(GE)^2$. Digit Concatenation provides an ability to continuously create new constants through evolution and incrementally improve the fitness of solutions by adapting individual constants at the digit level. Persistant Random Constants presents a variation on GP's ERC. ERC, the widely adopted mechanism for constant creation in GP, suffers from a number of flaws as outlined in Chapter 5. For Persistant Random Constants, similar to ERC, a large population of random constants is initially generated. However, unlike ERC these constants remain available for evolution to reintroduce to the population at any point. This facilitates the maintenance of a more diverse selection of constants. Constant creation through $(GE)^2$ using the Digit Concatenation method as a base in the meta-Grammar allowed the automatic biasing by the solution-grammar chromosome of useful constants.

9.2.5 *Identification of Diversity Being a Function of Potential in the Environment*

If application of EC to dynamic environments is to be successful, the legacy problem of static environments, where a population is encouraged to converge to an optimal solution, must be overcome. For dynamic environments, the maintenance of diversity is key. This diversity allows the population to provide wide coverage of the solution landscape as it changes over time. GE's genotype-to-phenotype mapping is an ideal solution to this problem, where in the genotype space, individuals can be dispersed across a wide spectrum with variable lengths and different genotypic values. These genotypes are then mapped into the phenotypic solution space. In Chapter 8, it was found that this mechanism equips GE with an ability to avoid convergence and furthermore, enables it to maintain phenotypic diversity in line with opportunities presented on the fitness landscape. Phenotypic diversity becomes a function of the fitness landscape, while the genotype is free to evolve along neutral networks.

9.2.6 *Identification of Two Levels of Evolvability in GE*

This book encountered the benefits of evolvability in two chapters. A high level of evolvability imbues the ability upon a population to incrementally evolve to better solutions, when it is in a local optima navigating to the global optimum or following an optima across a changing landscape. At the level of the BNF grammar in Chapter 5, Digit Concatenation was seen to achieve more accurate fitnesses because of its ability to incrementally evolve to the best solution. In Chapter 8, the population of solutions was seen to quickly adapt to new solutions after periods where the phenotypic diversity decreased due to the index making a loss. This speedy adaptation was brought about through GE's genotype-to-phenotype mapping, allowing it to continue evolutionary search through neutral mutations, while still maintaining phenotypic fitness.

9.2.7 *Experimental Evidence of the Evolution of Robust Solutions over Dynamic Data*

In Chapter 7, the weakness of evolving a population over a static data set was highlighted, where optimal solutions were produced but these solutions were brittle and overfit the training data. Chapter 8 provided evidence that where a population was progressively trained over dynamic data, the resulting solutions were more robust. On average in 50% of window increments the best-performing solution survived to trade again on the live data. Through being exposed to this dynamic data, GE was able to uncover more valuable underlying data that enabled it to produce solutions that transferred successfully to live or out-of-sample data.

9.2.8 Experimental Evidence of the Presence of Useful Information in Real-World Financial Historical Time Series

This book simulated live trading over historical financial time series. As a means of generating trading rules, a relatively simple grammar was used that included one technical indicator, the moving average indicator. However, when combined with GE, this grammar was able to produce rules that outperformed a benchmark buy-and-hold strategy in 23% of runs conducted on the Nikkei 225 data, while also producing competitive results on a strongly upward trending S&P 500 index. This implies for the Nikkei 225 that, for the period under investigation, the index did not exhibit efficient price formation when compared against the buy-and-hold benchmark, as the index was beaten through exclusive use of technical analysis on publicly available information. However, when considering the EMH as a joint hypothesis it becomes difficult to reject efficiency as the chosen benchmark may not be suitable. the use of financial data represents the introduction of a new real-world test-bed for the analysis of EC approaches to dynamic environments that experience complex types of change. Prior benchmarks focus on Markov or Deterministic types of change and, unlike these benchmarks, no standardised benchmark code is required, as the driver is widely available standardised data.

9.3 Opportunities for Future Research

This study represents an initial step in the exploration and application of GE to dynamic environments and as such represents a foundation for further application and development of GE in this area.

Chapter 3 outlined a number of approaches other researchers had adopted in dealing with dynamic environments. Memory, multiple populations, and problem decomposition were not explored in this book. Of these, there exists much scope for the exploration of the existence of memory within meta-Grammars and $(GE)^2$. For deterministic types of change, it may be possible for the solution-grammar chromosome to evolve useful biases and terminal subsets, allowing the solution chromosome to evolve a representation that allows efficient switching between solutions.

Further research opportunities in this area also exist in examining the potential of the solution-grammar chromosomes to evolve useful building blocks, which ties in with the idea of forming a memory in this chromosome. Useful sections of code or ADFs may be evolved in the solution-grammar chromosome, allowing the solution chromosome to then essentially conduct its evolutionary search with bigger pieces. Such an approach may enable $(GE)^2$ to remember *how* to produce fit solutions, not just fit solutions themselves. However, attention must be paid to maintaining levels of evolvability, which was demonstrated in this book to be of importance for dynamic environments.

A natural extension of this work is in replacing the standard GE mapping process with that of πGE. πGE has demonstrated an ability to produce superior results to those of standard GE in static problems. The application of πGE to dynamic problems may lead to an improvement in results of the same order.

This book conducted an examination of GE in the setting of a real-world problem that experienced complex change. Further application of EC paradigms to dynamic real-world problems may focus research on issues in those domains, rather than on limitations of catch-all benchmark problems. Exclusive research against benchmark problems runs the risk of tailoring algorithms to such problems.

9.4 Finally

Evolution in the natural world is an ongoing process that has led to a diverse array of complex living organisms that are capable of surviving in challenging dynamic environments. These organisms continue to evolve and survive as their environment changes or else fail to evolve and face extinction. As researchers, the utilisation of this process presents immense promise in providing dynamic solutions to the many real-world problems that experience some form of change with time. This book has provided a foundation on the road to realising that goal. The types of changes that can be faced by modellers were described. GE was then analysed in this context developing a greater understanding of evolvability, diversity and the quality of solutions evolved on dynamic data. This area represents a great opportunity for future research that must be conducted if the full potential of the evolutionary process is to be realised in the real-world, and this book represents just the initial foundations upon which future research can take place.

References

1. Abass, H.A., Sastry, K., Goldberg, D.E.: Oiling the wheels of change, the role of adaptive automatic problem decomposition in non-stationary environments, IlliGAL Report No. 2004029, University of Illinois at Urbana-Champaign (2004)
2. Allen, F., Karjalainen, R.: Using genetic algorithms to find technical trading rules. Journal of Financial Economics 51, 245–271 (1999)
3. Altenberg, L.: The evolution of evolvability in genetic programming. In: Kinnear Jr., K.E. (ed.) Advances in Genetic Programming, pp. 47–74. MIT Press, Cambridge (1994)
4. Andrews, M., Tuson, A.: Diversity does not necessarily imply adaptability. In: Proceedings of GECCO 2003, Workshop on Evolutionary Algorithms for Dynamic Optimization Problems, pp. 24–28 (2003)
5. Angeline, P.J.: Two Self-Adaptive Crossover Operators for Genetic Programming. In: Kinnear Jr., K.E. (ed.) Advances in Genetic Programming, vol. 2, pp. 89–110. MIT Press, Cambridge (1996)
6. Angeline, P.J.: Tracking extrema in dynamic environments. In: Angeline, P.J., McDonnell, J.R., Reynolds, R.G., Eberhart, R. (eds.) EP 1997. LNCS, vol. 1213, pp. 335–345. Springer, Heidelberg (1997)
7. Amarteifio, S.: Interpreting a Genotype-Phenotype Map with Rich Representations in XMLGE, Masters Thesis, University of Limerick (2005), `http://ncra.ucd.ie/downloads/pub/SaoirseMScThesis.pdf`
8. Bachelier, L.: Theorie de la speculation. Annales Scientifiques de l'Ecole Normale Superieure 3(17), 21–86 (1900)
9. Banzhaf, W.: Genotype-Phenotype Mapping and Neutral Variation - A case study in Genetic Programming. In: Davidor, Y., Männer, R., Schwefel, H.-P. (eds.) PPSN 1994. LNCS, vol. 866, pp. 322–332. Springer, Heidelberg (1994)
10. Banzhaf, W., Nordin, P., Keller, R.E., Francone, F.D.: Genetic Programming – An Introduction: On the Automatic Evolution of Computer Programs and its Applications. Morgan Kaufmann, San Francisco (1998)
11. Becker, L.A., Seshadri, M.: GP-evolved technical trading rules can outperform buy and hold. In: Proceedings of the Sixth International Conference on Computational Intelligence and Natural Computing, Embassy Suites Hotel and Conference Center, Cary, North Carolina USA, September 26-30, 2003, pp. 26–30 (2003)

12. Bendtsen, C.N., Krink, T.: Dynamic memory model for non-stationary optimization. In: Proceedings of IEEE Congress on Evolutionary Computation, pp. 145–150. IEEE Press, Los Alamitos (2002)
13. Blackwell, T.M.: Dynamic Search with Charged Swarms. In: Proceedings of 2002 Genetic and Evolutionary Computation Conference, GECCO 2002, pp. 19–26. Morgan Kaufmann, San Francisco (2002)
14. Bonabeau, E., Dorigo, M., Theraulaz, G.: Swarm Intelligence: From natural to artificial systems. Oxford University Press, Oxford (1999)
15. Brabazon, A., O'Neill, M., Ryan, C., Collins, J.J.: Uncovering Technical Trading Rules Using Evolutionary Automatic Programming. In: Proceedings of 2001 AAANZ Conference (Accounting Association of Australia and NZ) in Auckland, July 1-3 (2001)
16. Brabazon, A., O'Neill, M., Ryan, C., Matthews, R.: Evolving classifiers to model the relationships between strategy and corporate performance using grammatical evolution. In: Foster, J.A., Lutton, E., Miller, J., Ryan, C., Tettamanzi, A.G.B. (eds.) EuroGP 2002. LNCS, vol. 2278, pp. 103–113. Springer, Heidelberg (2002)
17. Brabazon, A., O'Neill, M.: Trading foreign exchange markets using evolutionary automatic programming. In: Proceedings of the Bird of a Feather Workshops, Genetic and Evolutionary Computation Conference, New York, pp. 133–136 (July 2002)
18. Brabazon, A., O'Neill, M.: A grammar model for foreign exchange trading. In: Proceedings of IC-AI 2003 the International Conference on Artificial Intelligence, pp. 492–499. CSEA Press (2003)
19. Brabazon, A., O'Neill, M.: Anticipating Bankruptcy Reorganisation from Raw Financial Data Using Grammatical Evolution. In: Raidl, G.R., Cagnoni, S., Cardalda, J.J.R., Corne, D.W., Gottlieb, J., Guillot, A., Hart, E., Johnson, C.G., Marchiori, E., Meyer, J.-A., Middendorf, M. (eds.) EvoIASP 2003, EvoWorkshops 2003, EvoSTIM 2003, EvoROB/EvoRobot 2003, EvoCOP 2003, EvoBIO 2003, and EvoMUSART 2003. LNCS, vol. 2611, pp. 368–378. Springer, Heidelberg (2003)
20. Brabazon, A., O'Neill, M.: Diagnosing Corporate Stability Using Grammatical Evolution. International Journal of Applied Mathematics and Computer Science 14(3), 363–374 (2004)
21. Brabazon, A., O'Neill, M.: Bond-issuer credit rating with grammatical evolution. In: Raidl, G.R., Cagnoni, S., Branke, J., Corne, D.W., Drechsler, R., Jin, Y., Johnson, C.G., Machado, P., Marchiori, E., Rothlauf, F., Smith, G.D., Squillero, G. (eds.) EvoWorkshops 2004. LNCS, vol. 3005, pp. 268–277. Springer, Heidelberg (2004)
22. Brabazon, A., O'Neill, M.: Evolving Technical Trading Rules for Spot Foreign-Exchange Markets Using Grammatical Evolution. Computational Management Science 1(3-4), 293–310 (2004)
23. Brabazon, A., O'Neill, M.: Credit Rating with piGrammatical Evolution. In: Proceedings of Computer Methods and Systems CMS 2005, Oprogramowanie Naukowo-Techniczne, Krakow, Poland, vol. 1, pp. 253–260 (2005)
24. Brabazon, A., O'Neill, M.: Biologically inspired algorithms for Financial Modelling. Springer, Heidelberg (2006)
25. Brabazon, A., O'Neill, M., Dempsey, I.: An Introduction to Evolutionary Computation in Finance. IEEE Computational Intelligence Magazine, 42–55 (November 2008)

26. Bradley, E., Tibshirani, R.: An introduction to the bootstrap. Chapman and Hall, Boca Raton (1994)
27. Branke, J.: Evolutionary approaches to dynamic environments - a survey. In: GECCO Workshop on Evolutionary Algorithms for Dynamic Optimization Problems, pp. 134–137 (1999)
28. Branke, J.: Evolutionary Algorithms for Dynamic Optimization Problems: A survey, Tech. Rep. 387, Intituge AIVB, University of Karlsruhe (February 1999)
29. Branke, J.: Memory enhanced evolutionary algorithms for changing optimization problems. In: Proceedings of IEEE Congress on Evolutionary Computation (CEC 1999), vol. 3, pp. 1875–1882. IEEE Press, Los Alamitos (1999)
30. Branke, J., Kausler, T., Schmidt, C., Schmek, H.: A multi-population approach to dynamic optimization problems. In: Adaptive Computing in Design and Manufacturing. Springer, Heidelberg (2000)
31. Branke, J.: Evolutionary approaches to dynamic environments - updated survey. In: GECCO Workshop on Evolutionary Algorithms for Dynamic Optimization Problems, pp. 27–30 (2001)
32. Branke, J.: Evolutionary Optimization in Dynamic Environments. Kluwer, Dordrecht (2001)
33. Branke, J., Salihoglu, E., Uyar, S.: Towards an Analysis of Dynamic Environments. In: Proceedings of the Genetic and Evolutionary Computation Conference (GECCO 2005), pp. 1433–1440. ACM Press, New York (2005)
34. Brock, W., Lokonishok, J., LeBaron, B.: Simple technical trading rules and the stochastic properties of stock returns. Journal of Finance 47(5), 1731–1764 (1992)
35. Brown, S., Goetzmann, W., Kumar, A.: The dow theory: William Peter Hamilton's track record reconsidered. Journal of Finance 53(4), 1311–1333 (1998)
36. Burke, E., Gustafson, S., Kendall, G.: A survey and analysis of diversity measures in genetic programming. In: Proceedings of the Genetic and Evolutionary Computation Conference, GECCO 2002, pp. 716–723. Morgan Kaufmann, San Francisco (2002)
37. Calabretta, R., Galbiati, R., Nolfi, S., Parisi, D.: Investigating the role of diploidy in simulated populations of evolving individuals. In: Electronic Proceedings of the 1997 European Conference on Artificial Life (1997)
38. Castillo, E., Gutierrez, J.: Nonlinear time series modeling and prediction using functional networks. Extracting information masked by chaos. Physics Letters A 244, 71–84 (1998)
39. Cedeno, W., Vemuri, V.R.: On the use of niching for dynamic landscapes. In: Proceedings of Intl. Conf. on Evolutionary Computation, pp. 361–366. IEEE Press, Los Alamitos (1997)
40. Chen, S., Yeh, C.: Toward a computable approach to the efficient market hypothesis: an application of genetic programming. Journal of Economic Dynamics and Control 21, 1043–1063 (1996)
41. Chomsky, N.: Reflections on Language. Pantheon Books, New York (1975)
42. Cleary, R., O'Neill, M.: An attribute grammar decoder for the (0,1) multiconstrained knapsack problem. In: Raidl, G.R., Gottlieb, J. (eds.) EvoCOP 2005. LNCS, vol. 3448, pp. 34–45. Springer, Heidelberg (2005)
43. Cleary, R.: Extending Grammatical Evolution with Attribute Grammars: An Application to Knapsack Problems, Masters Thesis, University of Limerick (2005), `http://ncra.ucd.ie/downloads/pub/thesisExtGEwithAGs-CRC.pdf`

44. Cobb, H.G.: An investigation into the use of hypermutation as an adaptive operator in genetic algorithms having continuous, time-dependent nonstationary environments, Technical Report AIC-90-001, Naval Research Laboratory, Washington, USA (1990)
45. Collingwood, E., Corne, D., Ross, P.: Useful Diversity via Multiploidy. In: IEEE International Conference on Evolutionary Computing, pp. 810–813. IEEE Press, Los Alamitos (1996)
46. Cramer, N.L.: A Representation for the Adaptive Generation of Simple Sequential Programs. In: Proceedings of the International Conference on Genetic Algorithms and Their Applications, Carnagie Mellon University, Pittsburgh, pp. 183–187. Lawrence Erlbaum Associates, Inc, Mahwah (1985)
47. The Credit Suisse/Tremont Hedge Fund index
48. Daida, J.M., Bertram, R.R., Stanhope, S.A., Khoo, J.C., Chaudhary, S.A., Chaudhri, O.A., Polito II, J.A.: What Makes a Problem GP-Hard? Analysis of a Tunably Difficult Problem in Genetic Programming. Genetic Programming and Evolvable Machines 2(2), 165–191 (2001)
49. Dasgupta, D., McGregor, D.R.: Nonstationary funciton optimization using the structured genetic algorithm. In: Manner, R., Maderick, B. (eds.) Proceedings of Parallel Problem Solving from Nature, pp. 145–154. Elsevier, Amsterdam (1992)
50. Dasgupta, D.: Incorporating redudancy and gene activation mechanisms in genetic search. In: Chambers, L. (ed.) Practical Handbook of Genetic Algorithms, vol. 2, pp. 303–316. CRC Press, Boca Raton (1995)
51. De Jong, K.A.: An analysis of the behavior of a class of genetic adaptive systems, Ph.D. thesis, University of Michigan (1975)
52. De Jong, K.A.: Evolving in a Changing World. In: Proceedings of the International Symposium on foundations of Intelligent Systems, pp. 512–519. Springer, Heidelberg (1999)
53. Dempsey, I., O'Neill, M., Brabazon, A.: Investigations into market index trading models using evolutionary automatic programming. In: O'Neill, M., Sutcliffe, R.F.E., Ryan, C., Eaton, M., Griffith, N.J.L. (eds.) AICS 2002. LNCS, vol. 2464, pp. 165–170. Springer, Heidelberg (2002)
54. Dempsey, I.: Grammatical Evolution in Dynamic Environments, PhD Thesis, University College Dublin, Ireland (2007)
55. Dempsey, I., O'Neill, M., Brabazon, A.: Constant Creation with Grammatical Evolution. International Journal of Innovative Computing and Applications 1(1), 23–38 (2007)
56. Dempsey, I., O'Neill, M., Brabazon, A.: Grammatical Constant Creation. In: Proceedings of the Genetic and Evolutionary Computation Conference, Part II, pp. 447–458. Springer, Heidelberg (2004)
57. Dempster, M.A.H., Jones, C.M.: A real-time adaptive trading system using genetic programming. Quantitative Finance 1, 397–413 (2001)
58. Dickmanns, D., Schmidhuber, J.: A. Winklhofer: Der genetische Algorithmus: Eine Implementierung in Prolog, Fortgeschrittenenpraktikum, Institut f. Informatik, Lehrstuhl Prof. Radig, Tech. Univ. Munich (1987)
59. Dynamic Optimisation Problems web-site, `http://www.aifb.uni-karlsruhe.de/~jbr/EvoDOP/`
60. Ebner, M., Shackelton, M., Shipman, R.: How neutral networks influence evolvability. Complexity 7(2), 19–33 (2001)

61. Eggermont, J., Lenaerts, T., Poyhonen, S., Termier, A.: Raising the dead: Extending evolutionary algorithms with a case-based memory. In: Miller, J.F., et al. (eds.) EuroGP 2001. LNCS, vol. 2038, pp. 280–290. Springer, Heidelberg (2001)
62. Eigen, M., McCaskill, J., Schuster, P.: The molecular quasispecies. Adv. Chem. Phys. 75, 149 (1989)
63. Evett, M., Fernandez, T.: Numeric Mutation Improves the Discovery of Numeric Constants in Genetic Programming. In: Proceedings of the Third Annual Conference on Genetic Programming, University of Wisconsin, Madison, Wisconsin, USA, pp. 66–71. Morgan Kaufmann, San Francisco (1998)
64. Fama, E.F.: Efficient Capital Markets: A Review of Theory and Empirical Work. Journal of Finance 25(2), 383–417 (1970)
65. Farrar, W.E.: Anti-Biotics Resistance in Developing Countries. Journal of Infectious Diseases 152, 1103–1106 (1985)
66. Feng, W., Brune, T., Chan, L., Chowdhury, M., Kuek, C.K., Li, Y.: Benchmarks for testing evolutionary algorithms. In: Proceedings of the Third Asia Pacific Conference on Measurement and Control, Dunhuang, China, pp. 134–138 (1998)
67. Fogel, D.B.: Evolutionary Computation: The Fossil Record. IEEE Press, Los Alamitos (1998)
68. Fogel, L., Owens, A.J., Walsh, M.J.: Artificial Intelligence Through Simulated Evolution. John Wiley, Chichester (1966)
69. Forrest, S., Mitchell, M.: Relative building-block fitness and the building-block hypothesis. In: Whitely, D. (ed.) Foundations of Genetic Algorithms, vol. 2. Morgan Kaufmann, San Francisco (1993)
70. de Franca, F.O., Von Zuben, F.J., de Castro, L.N.: An artificial immune network for multimodal function optimization on dynamic environments. In: Proceedings of the IEEE Congress on Evolutionary Computation 2005, vol. 1, pp. 289–296. ACM Press, New York (2005)
71. Friedberg, R.M.: A learning machine: Part I. IBM Journal (January 1958)
72. Friedberg, R.M., Dunham, B., North, J.H.: A learning machine: Part II. IBM Journal (July 1959)
73. Fujiki, C.: An evaluation of Holland's genetic operators applied to a program generator, Master's thesis, University of Idaho, Moscow, ID (1986)
74. Gaspar, A., Collard, P.: From Gas to Artificial Immune Systems: Improving Adaptation in Time Dependent Optimization. In: Proceedings of IEEE Congress on Evolutionary Conputation, Washington, pp. 1867–1874. IEEE Press, Los Alamitos (1999)
75. Genetic Programming web-site, `http://www.genetic-programming.com`
76. GEVA: Grammatical Evolution in Java, `http://ncra.ucd.ie/geva/`
77. Ghosh, A., Tstutsui, S., Tanaka, H.: Function optimization in nonstationary environment using steady state genetic algorithms with aging of individuals. In: Proceedings of IEEE Conference on Evolutinary Computation, pp. 666–671. IEEE Press, Los Alamitos (1998)
78. Goldberg, D.E., Smith, R.E.: Nonstationary function optimization using genetic algorithms with dominance and diploidy. In: Grefenstette, J.J. (ed.) International Conference on Genetic Algorithms, pp. 59–68. Lawrence Erlbaum Associates, Mahwah (1987)
79. Goldberg, D.E.: Genetic algorithms in search, optimization and machine learning. Addison Wesley, Reading (1989)

80. Goldberg, D., Sastry, K., Llora, X.: Toward routine billion-variable optimization using genetic algorithms. Complexity 12(3), 27–29 (2007)
81. Greffenstette, J.J.: Genetic Algorithms for Changing Environments. In: Proceedings of Parallel Problem solving from Nature, vol. 2, pp. 137–144. Elsevier, Amsterdam (1992)
82. Hadad, B.S., Eick, C.F.: Supporting polyploidy in genetic algorithms using dominance vectors. In: Angeline, P.J., et al. (eds.) EP 1997. LNCS, vol. 1213, pp. 223–234. Springer, Heidelberg (1997)
83. Hansen, T.F.: Is modularity necessary for evolvability? Remarks on the relationship between pleiotropy and evolvability. BioSystems 69, 83–94 (2002)
84. Hansen, J.V., Lowry, P.B., Meservy, R.D., McDonald, D.M.: Genetic Programming for Prevention of Cyberterrorism through dynamic and evolving intrusion detection. Decision Support Systems 43(4), 1362–1374 (2006)
85. Harper, R., Blair, A.: A structure preserving crossover in grammatical evolution. In: Proceedings of the IEEE Congress on Evolutionary Computation (CEC 2005), pp. 2537–2544. IEEE Press, Los Alamitos (2005)
86. Hemberg, E., Gilligan, C., O'Neill, M., Brabazon, A.: A Grammatical Genetic Programming Approach to Modularity in Genetic Algorithms. In: Ebner, M., O'Neill, M., Ekárt, A., Vanneschi, L., Esparcia-Alcázar, A.I. (eds.) EuroGP 2007. LNCS, vol. 4445, pp. 1–11. Springer, Heidelberg (2007)
87. Hemberg, M., O'Reilly, U.-M.: GENR8 - Extending Grammatical Evolution to Evolve Digital Surfaces with Genr8. In: Keijzer, M., O'Reilly, U.-M., Lucas, S., Costa, E., Soule, T. (eds.) EuroGP 2004. LNCS, vol. 3003, pp. 299–309. Springer, Heidelberg (2004)
88. Hicklin, J.: Application of the genetic algorithm to automatic program generation, Master's thesis, University of Idaho, Moscow, ID (1986)
89. Holland, J.H.: Adaptation in natural and artificial systems. MIT Press, Cambridge (1975)
90. Holland, J.H.: Adaption in Natural and Artificial Systems. University of Michigan Press (1980)
91. Holland, J.H.: Emergence from Chaos to Order. Oxford University Press, Oxford (1998)
92. Hu, J., Goodman, E.: Topological Synthesis of Robust Dynamic Systems for Sustainable Genetic Programming. In: Genetic Programming Theory and Practice, pp. 143–157. Springer, Heidelberg (2004)
93. Huang, C., Rocha, L.: Exploration of RNA editing and design of robust genetic algorithms. In: Proceedings of the 2003 IEEE Congress on Evolutionary Computation, pp. 2799–2806. IEEE Press, Los Alamitos (2003)
94. Huang, C., Rocha, L.: A systematic study of genetic algorithms with genotype editing. In: Proceedings of the 2004 Genetic and Evolutionary Computation Conference, vol. 1, pp. 1233–1245. Springer, Heidelberg (2004)
95. Huang, C., Rocha, L.R.: Tracking Extrema in Dynamic Environments using a Coevolutionary Agent-based Model of Genotype Edition. In: Proceedings of the Genetic and Evolutionary Computation Conference (GECCO 2005), pp. 545–552. ACM Press, New York (2005)
96. Iba, H., Nikolaev, N.: Genetic programming polynomial models of financial data series. In: Proceedings of the IEEE Congress on Evolutionary Computation (CEC 2000), pp. 1459–1466. IEEE Press, Los Alamitos (2000)
97. Jakobović, D., Budin, L.: Dynamic scheduling with genetic programming. In: Collet, P., Tomassini, M., Ebner, M., Gustafson, S., Ekárt, A. (eds.) EuroGP 2006. LNCS, vol. 3905, pp. 73–84. Springer, Heidelberg (2006)

98. Kahneman, D., Tversky, A.: Prospect theory: an analysis of decision under risk. Econometrica 47, 263–291 (1979)
99. Kavajecz, K.A., Odders-White, E.R.: Technical analysis and liquidity provision. Review of Financial Studies 17(4), 1043–1071 (2004)
100. Karaman, A., Uyar, A.Ş., Eryiğit, G.: The memory indexing evolutionary algorithm for dynamic environments. In: Rothlauf, F., Branke, J., Cagnoni, S., Corne, D.W., Drechsler, R., Jin, Y., Machado, P., Marchiori, E., Romero, J., Smith, G.D., Squillero, G. (eds.) EvoWorkshops 2005. LNCS, vol. 3449, pp. 563–573. Springer, Heidelberg (2005)
101. Keijzer, M.: Efficiently representing populations in genetic programming. In: Angeline, P., Kinnear Jr., K. (eds.) Advances in Genetic Programming, vol. 2, pp. 259–278. MIT Press, Cambridge (1996)
102. Keijzer, M.: Improving Symbolic Regression with Interval Arithmetic and Linear Scaling. In: Ryan, C., Soule, T., Keijzer, M., Tsang, E.P.K., Poli, R., Costa, E. (eds.) EuroGP 2003. LNCS, vol. 2610, pp. 70–82. Springer, Heidelberg (2003)
103. Keller, R., Banzhaf, W.: The evolution of genetic code in genetic programming. In: Proceedings of the Genetic and Evolutionary Computation Conference, Orlando Florida, USA, vol. 2, pp. 1077–1082. Morgan Kaufman, San Francisco (1999)
104. Kendall, M.G., Bradford Hill, A.: The analysis of economic time-series-part 1: prices. Journal of the Royal Statistical Society Series A (General) 116(1), 11–34 (1953)
105. Kennedy, J., Eberhart, R., Shi, Y.: Swarm Intelligence. Morgan Kaufmann, San Francisco (2001)
106. Kennedy, J., Eberhart, R.: Particle swarm optimization. In: Proceedings of the IEEE International Conference on Neural Networks, pp. 1942–1948. IEEE Press, Los Alamitos (1995)
107. Kibria, R.H., Li, Y.: Optimizing the initialization of dynamic decision heuristics in DPLL SAT solvers using genetic programming. In: Collet, P., Tomassini, M., Ebner, M., Gustafson, S., Ekárt, A. (eds.) EuroGP 2006. LNCS, vol. 3905, pp. 331–340. Springer, Heidelberg (2006)
108. Kimura, M.: The Neutral Theory of Evolution. Cambridge University Press, Cambridge (1983)
109. Kolorenc, J.: Evolving Phonological Rules Using Grammatical Evolution. In: Prodceedings of the 8th International Student Confernce on Electrical Engineering 2004, Prague, Czech Republic (May 2004)
110. Koza, J.R.: Genetic Programming: On the Programming of Computers by Means of Natural Selection. MIT Press, Cambridge (1992)
111. Koza, J.R.: Genetic Programming II: Automatic Discovery of Reusable Programs. MIT Press, Cambridge (1994)
112. Koza, J.R., Andre, D., Bennett III, F.H., Keane, M.: Genetic Programming III: Darwinian Invention and Problem Solving. Morgan Kaufmann, San Francisco (1999)
113. Koza, J.R., Keane, M., Streeter, M.J., Mydlowec, W., Yu, J., Lanza, G.: Genetic Programming IV: Routine Human-Competitive Machine Intelligence. Kluwer Academic Publishers, Dordrecht (2003)
114. Kwon, Y., Moon, B.: Evolutionary ensemble for stock prediction. In: Deb, K., et al. (eds.) GECCO 2004. LNCS, vol. 3103, pp. 1102–1113. Springer, Heidelberg (2004)

115. Langdon, W.B., Gustafson, S., Koza, J.R.: The Genetic Programming Bibliography, `http://www.cs.bham.ac.uk/~wbl/biblio/`
116. Langdon, W.B., Poli, R.: Genetic programming bloat with dynamic fitness. In: Banzhaf, W., Poli, R., Schoenauer, M., Fogarty, T.C. (eds.) EuroGP 1998. LNCS, vol. 1391, pp. 96–112. Springer, Heidelberg (1998)
117. Lenoski, R.E.: Bacterial Evolution and the Cost of Anti-Biotic Resistance. International Microbiology 1, 265–270 (1998)
118. Lewin, B.: Genes VII. Oxford University Press (2000)
119. Lewis, J., Hart, E., Ritchie, G.: A comparison of dominance mechanisms and simple mutation on non-stationary problems. In: Eiben, A.E., Bäck, T., Schoenauer, M., Schwefel, H.-P. (eds.) PPSN 1998. LNCS, vol. 1498, pp. 139–148. Springer, Heidelberg (1998)
120. Li, J., Tsang, E.P.K.: Improving technical analysis predictions: An application of genetic programming. In: Proceedings of Florida Artificial Intelligence Research Symposium, USA (1999)
121. Li, X., Branke, J., Blackwell, T.: Particle Swarm with Speciation Adaptation in a Dynamic Environment. In: Proceedings of the Genetic and Evolutionary Colmputation Conference (GECCO 2006), pp. 51–58. ACM Press, New York (2006)
122. Lo, A.: Market Efficiency: Stock Market Behaviour in Theory and Practice. Edward Elgar Publishing Ltd (1997)
123. Lo, A., Mamaysky, H., Wang, J.: Foundations of technical analysis: computational algorithms, statistical interference, and empirical implementation. Journal of Finance 55(4), 1705–1765 (2000)
124. Lo, A.: The Adaptive Market Hypothesis: market efficiency from an evolutionary perspective. Journal of Portfolio Managment 30, 15–29 (2004)
125. Lo, A.: Reconciling efficient markets with behavioral finance: the adaptive market hypothesis. Journal of Investment Consulting 7(2), 21–43 (2005)
126. Louis, S.J., Zu, Z.: Genetic algorithms for open shop scheduling and rescheduling. In: Cohen, M.E., Hudson, D.L. (eds.) ISCA 11th Intl. Conf. on Computers and their Applications, pp. 99–102 (1996)
127. Luijten, S.H., Kery, M., Gerrard, J., Oostermeijer, B., Den Nijs, H.C.M.: Demographic consequences of inbreeding and outbreeding in Arnica montana: a field experiment. Journal of Ecology 90(4), 593 (2002)
128. Mauldin, M.: Maintaining Diversity in Genetic Search. In: Proceedings of the National Conference on Artificial Intelligence, pp. 247–250 (August 1984)
129. May, R.: Simple mathematical models with very complicated dynamics. Nature 261, 459–467 (1976)
130. Moore, J., Hahn, L.: Grammatical evolution for the discovery of petri net models of complex genetic systems. In: Cantú-Paz, E., Foster, J.A., Deb, K., Davis, L., Roy, R., O'Reilly, U.-M., Beyer, H.-G., Kendall, G., Wilson, S.W., Harman, M., Wegener, J., Dasgupta, D., Potter, M.A., Schultz, A., Dowsland, K.A., Jonoska, N., Miller, J., Standish, R.K. (eds.) GECCO 2003. LNCS, vol. 2724, pp. 2412–2413. Springer, Heidelberg (2003)
131. Mori, N., Kita, H., Nishikawa, Y.: Adaptation to a changing environment by means of the thermodynamical genetic algorithm. In: Babaoğlu, Ö., Marzullo, K. (eds.) WDAG 1996. LNCS, vol. 1151, pp. 513–522. Springer, Heidelberg (1996)

132. Mori, N., Imanishi, S., Kita, H., Nishikawa, Y.: Adaptation to changing environments by means of the memory based thermodynamical genetic algorithm. In: Proceedings of the International Conference on Genetic Algorithms, pp. 299–306. Morgan Kaufmann, San Francisco (1997)
133. Mori, N., Kita, H., Nishikawa, Y.: Adaptation to a changing environment by means of the feedback thermodynamical genetic algorithm. In: Eiben, A.E., Bäck, T., Schoenauer, M., Schwefel, H.-P. (eds.) PPSN 1998. LNCS, vol. 1498, pp. 149–157. Springer, Heidelberg (1998)
134. Morrison, R.W., DeJong, K.A.: A test problem generator for non-stationary environments. In: Proceedings of the International Congress on Evolutionary Computation, vol. 3, pp. 2047–2053. IEEE Computer Society Press, Los Alamitos (1999)
135. Morrison, R.W.: Designing Evolutionary Algorithms for Dynamic Environments. Springer, Heidelberg (2004)
136. Morrison, R.W.: Dispersion-based population initialisation. In: Cantú-Paz, E., Foster, J.A., Deb, K., Davis, L., Roy, R., O'Reilly, U.-M., Beyer, H.-G., Kendall, G., Wilson, S.W., Harman, M., Wegener, J., Dasgupta, D., Potter, M.A., Schultz, A., Dowsland, K.A., Jonoska, N., Miller, J., Standish, R.K. (eds.) GECCO 2003. LNCS, vol. 2724, pp. 1210–1221. Springer, Heidelberg (2003)
137. Motsinger, A.A., Reif, D.M., Dudek, S.M., Ritchie, M.D.: Understanding the Evolutionary Process of Grammatical Evolution Neural Networks for Feature Selection in Genetic Epidemiology. In: Proceedings of the IEEE Symposium on Computational Intelligence and Bioinformatics and Computational Biology, pp. 1–8. IEEE Press, Los Alamitos (2006)
138. Murphy, J.J.: Technical analysis of the financial markets, New York Institute of Finance (1999)
139. Neely, C., Weller, P., Dittmar, R.: Is technical analysis in the foreign exchange market profitable? A genetic programming approach. Journal of Financial and Quantitative Analysis 32(4), 405–426 (1997)
140. Newman, M.E.J., Engelhart, R.: Effects of neutral selection on the evolution of molecular species. Proc. R. Soc., London B 256, 1333–1338 (1998)
141. Nicolau, M., Dempsey, I.: Grammar Based Extensions for Grammatical Evolution. In: Proceedings of the Congress on Evolutionary Computation 2006, pp. 2589–2592. IEEE Press, Los Alamitos (2006)
142. Nie, J.: Nonlinear time-series forecasting: A fuzzy-neural approach. Neurocomputing 16, 63–76 (1997)
143. Nikolaev, N., Iba, H.: Regularization Approach to Inductive Genetic Programming. IEEE Transactions on Evolutionary Computing 54(4), 359–375 (2001)
144. Ng, K.P., Wong, K.C.: A new diploid scheme and dominance change mechanism for non-stationary function optimization. In: 6th Intl. Conf. on Genetic Algorithms, pp. 159–166. Morgan Kaufmann, San Francisco (1995)
145. O'Neill, M., Ryan, C.: Automatic Generation of Caching Algorithms. In: Proceedings of EUROGEN 1999, Finland, pp. 127–134. University of Jyväskylä (1999)
146. O'Neill, M., Collins, J.J., Ryan, C.: Automatic Programming of Robots. In: Proceedings of Irish Conference in Artificial Intelligence and Cognitive Science 2000 (2000)

147. O'Neill, M., Collins, J.J., Ryan, C.: Automatic Generation of Robot Behaviours using Grammatical Evolution. In: Proceedings of AROB 2000, the Fifth International Symposium on Artificial Life and Robotics, pp. 351–354 (2000)
148. O'Neill, M.: Automatic Programming in an Arbitrary Language: Evolving Programs in Grammatical Evolution, PhD thesis, University of Limerick (2001)
149. O'Neill, M., Brabazon, A., Ryan, C., Collins, J.: Developing a market timing system using grammatical evolution. In: Proceedings of the Genetic and Evolutionary Computation Conference (GECCO 2001), pp. 1375–1381. Morgan Kaufmann, San Francisco (2001)
150. O'Neill, M., Brabazon, A., Ryan, C., Collins, J.J.: Evolving market index trading rules using grammatical evolution. In: Boers, E.J.W., Gottlieb, J., Lanzi, P.L., Smith, R.E., Cagnoni, S., Hart, E., Raidl, G.R., Tijink, H. (eds.) EvoIASP 2001, EvoWorkshops 2001, EvoFlight 2001, EvoSTIM 2001, EvoCOP 2001, and EvoLearn 2001. LNCS, vol. 2037, pp. 343–352. Springer, Heidelberg (2001)
151. O'Neill, M., Ryan, C.: Grammatical Evolution. IEEE Trans. Evolutionary Computation 5(4), 349–358 (2001)
152. O'Neill, M., Brabazon, A., Ryan, C.: Forecasting market indices using evolutionary automatic programming: A case study. In: Chen, S.-H. (ed.) Genetic Algorithms and Genetic Programming in Economics and Finance, pp. 174–195. Kluwer Academic Publishers, Dordrecht (2002)
153. O'Neill, M., Dempsey, I., Brabazon, A., Ryan, C.: Analysis of a Digit Concatenation Approach to Constant Creation. In: Ryan, C., Soule, T., Keijzer, M., Tsang, E.P.K., Poli, R., Costa, E. (eds.) EuroGP 2003. LNCS, vol. 2610, pp. 173–182. Springer, Heidelberg (2003)
154. O'Neill, M., Ryan, C.: Grammatical Evolution: Evolutionary Automatic Programming in an Arbitrary Language. Kluwer Academic Publishers, Dordrecht (2003)
155. O'Neill, M., Ryan, C., Keijzer, M., Cattolico, M.: Crossover in Grammatical Evolution. Genetic Programming and Evolvable Machines 4(1), 67–93 (2003)
156. O'Neill, M., Ryan, C.: Grammatical Evolution by Grammatical Evolution: The Evolution of Grammar and Genetic Code. In: Keijzer, M., O'Reilly, U.-M., Lucas, S., Costa, E., Soule, T. (eds.) EuroGP 2004. LNCS, vol. 3003, pp. 138–149. Springer, Heidelberg (2004)
157. O'Neill, M., Brabazon, A., Nicolau, M., McGarraghy, S., Keenan, P.: π Grammatical Evolution. In: Deb, K., et al. (eds.) GECCO 2004. LNCS, vol. 3103, pp. 617–629. Springer, Heidelberg (2004)
158. O'Neill, M., Cleary, R., Nikolov, N.: Solving knapsack problems with attribute grammars. In: Grammatical Evolution Workshop 2004. Proceedings of the Workshops, Genetic and Evolutionary Computation Conference (GECCO 2004), Seattle, WA, USA (June 2004)
159. O'Neill, M., Brabazon, A.: mGGA: the meta-Grammar genetic algorithm. In: Keijzer, M., Tettamanzi, A.G.B., Collet, P., van Hemert, J., Tomassini, M. (eds.) EuroGP 2005. LNCS, vol. 3447, pp. 311–320. Springer, Heidelberg (2005)
160. O'Neill, M., Brabazon, A.: Recent Adventures in Grammatical Evolution. In: Proceedings of Computer Methods and Systems conference (CMS 2005), Krakow, Poland, vol. 1, pp. 245–253 (2005)

161. O'Neill, M., Adley, C., Brabazon, A.: A Grammatical Evolution Approach to Eukaryotic Promoter Recognition. In: Proceedings of Bioinformatics Inform Workshop and Symposium, Dublin City University, Dublin (2005)
162. O'Neill, M., Brabazon, A.: Grammatical Differential Evolution. In: Proceedings of the 2006 International Conference on Artificial Intelligence (ICAI 2006), vol. 1, pp. 231–236. CSEA Press (2006)
163. O'Neill, M., Brabazon, A.: Grammatical Swarm: The Generation of Programs by Social Programming. Natural Computing 5(4), 443–462 (2006)
164. O'Neill, M., Brabazon, A.: Evolving a Logo Design using Lindenmayer Systems, Postscript and Grammatical Evolution. In: Proceedings of the IEEE World Congress on Evolutionary Computation, pp. 3788–3794. IEEE Computer Society Press, Los Alamitos (2008)
165. Ohkura, K., Ueda, K.: Adaptation in dynamic environment using genetic algorithms with redundant representation and additional genetic operators. In: Proceedings of Intelligent Engineering Systems Through Artificial Neural Networks, pp. 291–296. Amer. Society of Mechanical Engineers (1994)
166. Ohkura, K., Ueda, K.: A genetic algorithm with neutral mutations for solvingnonstationary function optimization problems. In: Proceedings of the 1994 Second Australian and New Zealand Conference on Intelligent Information Systems, pp. 248–252. IEEE Press, Los Alamitos (1994)
167. Oppacher, F., Wineberg, M.: The shifting balance genetic algorithm: Improving the GA in a dynamic environment. In: Proceedings of Genetic and Evolutionary Computation Conference (GECCO 1999), vol. 1, pp. 504–510. Morgan Kaufmann, San Francisco (1999)
168. Ortega, A., Alfonoso, R., Moreno, M.: Automatic composition of music by means of grammatical evolution. In: Proceedings of the International Conference on APL: array processing languages: lore, problems, and applications, pp. 148–155. ACM Press, New York (2002)
169. Ortega, A., Dalhoum, A., Alfonseca, M.: Grammatical evolution to design fractal curves with a given dimension. IBM Journal of Research and Development 47(4), 483–494 (2003)
170. Osmera, P., Kvasnicka, V., Pospichal, J.: Genetic algorithms with diploid chromosomes. In: Proceedings of Mendel 1997, pp. 111–116 (1997)
171. Petrowski, A., Genet, M.G.: A classification tree for speciation. In: Proceedings of the IEEE Conference on Evolutionary Computation (CEC 1999), pp. 204–211. IEEE Press, Los Alamitos (1999)
172. Pinker, S.: The language instinct: the new science of language and the mind. Penguin (1995)
173. Poli, R., Langdon, W.B., McPhee, N.F.: A field guide to genetic programming (2008), `http://lulu.com`, `http://www.gp-field-guide.org.uk`
174. Potvin, J., Soriano, P., Vallee, M.: Generating trading rules on the stock markets with genetic programming. Computers & Operations Research 31, 1030–1047 (2004)
175. Price, K.: An introduction to differential evolution. In: Corne, D., Dorigo, M., Glover, F. (eds.) New Ideas in Optimization, pp. 79–108. McGraw-Hill, New York (1999)
176. Pring, M.: Technical analysis explained: the successful investor's guide to spotting investment trends and turning points. McGraw-Hill, New York (1991)

177. Ramsey, C.L., Grefenstette, J.J.: Case-based intialization of genetic algorithms. In: Proceedings of the 5th Intl. Conf. on Genetic Algorithms, pp. 84–91. Morgan Kaufmann, San Francisco (1993)
178. Rand, W., Riolo, R.: The Problem with a Self-Adaptive Mutation Rate in Some Environments. In: Proceedings of the Genetic and Evolutionary Computation Conference (GECCO 2005), pp. 1493–1500. ACM Press, New York (2005)
179. Reisinger, J., Stanley, K.O., Miikkulainen, R.: Towards an empirical measure of evolvability. In: Proceedings of the Genetic and Evolutionary Computation Conference (GECCO 2005), pp. 257–264. ACM Press, New York (2005)
180. Rissanen, J.J.: Modelling by shortest data description. Automatica 14, 465–471 (1978)
181. Rocha, L.: Contextual genetic algorithms: Evolving developmental rules. In: Morán, F., Merelo, J.J., Moreno, A., Chacon, P. (eds.) ECAL 1995. LNCS, vol. 929, pp. 368–382. Springer, Heidelberg (1995)
182. Rocha, L., Huang, C.: The role of RNA editing in dynamic environments. In: Proceedings of the Ninth International Conference on the Simulation and Syntesis of Living Systems (ALIFE9), pp. 489–494. MIT Press, Cambridge (2004)
183. Ronnewinkel, C., Wilke, C.O., Martinez, T.: Genetic algorithms in time-dependent environments. In: Proceedings of the 2nd EvoNet Summer School (1999)
184. Ronnewinkel, C., Martinez, T.: Explicit speciation with few a priori parameters for dynamic optimization problems. In: Proceedings of GECCO Workshop on Evolutionary Algorithms for Dynamic Optimization Problems, pp. 31–34 (2001)
185. Rosca, J.: Entropy-driven adaptive representation. In: Rosca, J. (ed.) Proceedings of the Workshop on Genetic Programming: From Theory to Real-World Applications, Tahoe City, California, USA, pp. 23–32 (1995)
186. Rothlauf, F., Oetzel, M.: On the locality of grammatical evolution. In: Collet, P., Tomassini, M., Ebner, M., Gustafson, S., Ekárt, A. (eds.) EuroGP 2006. LNCS, vol. 3905, pp. 320–330. Springer, Heidelberg (2006)
187. Russo, J., Shoemaker, P.: Decision traps: ten barriers to brilliant decision making and how to overcome them. Simon & Schuster (1989)
188. Ryan, C.: Pygmies and civil servants. In: Advances in Genetic Programming, pp. 243–263. MIT Press, Cambridge (1994)
189. Ryan, C.: Diploidy without dominance. In: 3rd Nordic Workshop on Genetic Algorithms, pp. 63–70 (1997)
190. Ryan, C., Collins, J.J., Neill, M.O.: Grammatical Evolution: Evolving Programs for an Arbitrary Language. In: Banzhaf, W., Poli, R., Schoenauer, M., Fogarty, T.C. (eds.) EuroGP 1998. LNCS, vol. 1391, pp. 83–95. Springer, Heidelberg (1998)
191. Ryan, C., Collins, J.J.: Polygenic inheritance - A haploid scheme that can outperform diploidy. In: Eiben, A.E., Bäck, T., Schoenauer, M., Schwefel, H.-P. (eds.) PPSN 1998. LNCS, vol. 1498, pp. 178–187. Springer, Heidelberg (1998)
192. Ryan, C., O'Neill, M., Collins, J.J.: Grammatical Evolution: Solving Trigonometric Identities. In: Proceedings of Mendel 1998: 4th International Mendel Conference on Genetic Algorithms, Optimisation Problems, Fuzzy Logic, Neural Networks, Rough Sets, Brno, Czech Republic, June 24-26, 1998, pp. 111–119 (1998)

193. Ryan, C., O'Neill, M.: How to do anything with Grammars. In: Proceedings of the Bird of a Feather Workshops, Genetic and Evolutionary Computation Conference 2002, pp. 116–119 (2002)
194. Ryan, C., Keijzer, M.: An Analysis of Diversity of Constants of Genetic Programming. In: Ryan, C., Soule, T., Keijzer, M., Tsang, E.P.K., Poli, R., Costa, E. (eds.) EuroGP 2003. LNCS, vol. 2610, pp. 404–413. Springer, Heidelberg (2003)
195. Saxen, H.: On the approximation of a quadratic map by a small neural network. Neurocomputing 12, 313–326 (1996)
196. Schonemann, L.: The impact of population sizes and diversity on the adaptability of evolution strategies in dynamic environments. In: Proceedings of the IEEE Congress on Evolutionary Computation, vol. 2, pp. 1270–1277. IEEE Press, Los Alamitos (2004)
197. Schoreels, C., Garibaldi, J.M.: The effect of varying parameters on performance for adaptive agents in technical equity market trading. In: Proceedings of the 3rd International Conference on Computational Cybernetics (ICCC 2005), pp. 243–248. IEEE Press, Los Alamitos (2005)
198. Sebag, M., Schoenaur, M., Ravise, C.: Toward civilised evolution: developing inhibitions. In: Proceedings of the 7th Intl. Conf. on Genetic Algorithms - VII ICGA 1997, pp. 291–298. Morgan Kaufmann, San Francisco (1997)
199. Sharpe, W.F.: The Sharpe Ratio. Journal of Portfolio Management, 49–58 (Fall 1994)
200. Shefrin, H.: Beyond Greed and Fear: Understanding behavioral finance and the psychology of investing. Oxford University Press, Oxford (2002)
201. Shipman, R., Shackleton, M., Harvey, I.: The use of neutral genotype-phenotype mappings for improved evolutionary search. BT Technical Journal 18(4), 103–111 (2000)
202. Silva, A., Neves, A., Costa, E.: An Empirical Comparison of Particle Swarm and Predator Prey Optimisation. In: O'Neill, M., Sutcliffe, R.F.E., Ryan, C., Eaton, M., Griffith, N.J.L. (eds.) AICS 2002. LNCS (LNAI), vol. 2464, pp. 103–110. Springer, Heidelberg (2002)
203. Simon, H.: A behavioural model of rational choice. Quarterly Journal of Economics 69, 99–118 (1955)
204. Simoes, A., Costa, E.: Using GAs to deal with dynamic environments: a comparative study of several approaches based on promoting diversity. In: Proceedings of the Genetic and Evolutionary Computation Conference (GECCO 2002), p. 698. Morgan Kaufmann, San Francisco (2002)
205. Simoes, A., Costa, E.: An immune system-based genetic algorithm to deal with dynamic environments: Diversity and memory. In: Proceedings of the Sixth international conference on neural networks and genetic algorithms (ICANNGA 2003), pp. 168–174. Springer, Heidelberg (2003)
206. Smith, R.E.: Diploid genetic algorithms for search in time varying environments. In: Annual Southeast Regional Conference of the ACM, pp. 175–180. ACM Press, New York (1987)
207. Spencer, G.: Automatic Generation of Programs for Crawling and Walking. In: Advances in Genetic Programming, pp. 335–353. MIT Press, Cambridge (1994)
208. Stephens, C., Waelbroeck, H.: Schemata evolution and building blocks. Evolutionary Computation 7(2), 109–124 (1999)

209. Storn, R., Price, K.: Differential evolution-a simple and efficient adaptive scheme for global optimization over continuous spaces, Technical Report TR-95-012: International Computer Science Institute, Berkeley (1995)
210. Storn, R., Price, K.: Differential evolution-a simple and efficient heuristic for global optimization over continuous spaces. Journal of Global Optimization 11, 341–359 (1997)
211. Storn, R.: System design by constraint adaptation and differential evolution. IEEE Transactions on Evolutionary Computation 3, 22–34 (1999)
212. Taylor, M., Allen, H.: The use of technical analysis in the foreign exchange market. Journal of International Money and Finance 11, 304–314 (1992)
213. Thaler, R.: The Winner's Curse: Paradoxes and Anomalies of Economic Life. Free Press (1991)
214. Trojanowski, K., Michalewicz, Z., Xiao, J.: Adding memory to the evolutionary planner/navigator. In: Proceedings of the IEEE Intl. Conference on Evolutionary Computation, pp. 483–487. IEEE Press, Los Alamitos (1997)
215. Trojanowski, K., Michalewicz, Z.: Evolutionary Algorithms for Non-Stationary Environments. In: Proceedings of the 8th Workshop: Intelligent Information Systems VIII, Ustron, Poland, June 14-18, 1999, pp. 229–240. ICS PAS Press (1999)
216. Trojanowski, K., Michalewicz, Z.: Evolutionary optimization in non-stationary environments. Journal of Computer Science and Technology 1(2), 93–124 (2000)
217. Tsang, E., Yung, P., Li, J.: EDDIE-Automation, a decision support tool of financial forecasting. Decision Support Systems 37(4), 559–565 (2004)
218. Tsutsui, S., Fujimoto, Y., Gosh, A.: Forking genetic algorithms: GAs with search space division schemes. Evolutionary Computation 5(1), 61–80 (1997)
219. Turing, A.: Intelligent Machinery, p. 12 (1948)
220. Turing, A.: Computing Machinery and Intelligence, p. 456 (1950)
221. Tversky, A., Kahneman, D.: Rational Choice and the Framing of Decisions. Journal of Business 59(S)4, S251–S278 (1986)
222. Ursem, R.K.: Multinational GAs: multimodal optimisation techniques in dynamic environments. In: Proceedings of Genetic and Evolutionary Computation Conference (GECCO 2000), pp. 19–26. Morgan Kaufmann, San Francisco (2000)
223. Uyar, A.S., Harmanci, A.E.: Performance comparisons of genotype-to-phenotype mapping schemes for diploid representations in changing environments. In: Proceedings of the International Conference on Recent Advances in Soft Computing, Nottingham, United Kingdom, pp. 128–134 (2002)
224. Uyar, A.S., Harmanci, A.E.: A new population based adaptive domination change mechanism for diploid genetic algorithms in dynamic environments. Soft Computing - A Fusion of Foundations, Methodologies and Applications 9(11), 803–814 (2005)
225. Vavak, F., Jukes, K., Fogarty, T.C.: Adaptive combustion balancing in a multiple burner boiler using a genetic algorithm with variable local search. In: Proceedings of the Seventh Intl. Conf. on Genetic Algorithms, pp. 602–608. Morgan Kaufmann, San Francisco (1997)
226. Vavak, F., Jukes, K., Fogarty, T.C.: Learning the local search range for genetic optimisation in nonstationary environments. In: Proceedings of the IEEE Intl. Conf. on Evolutionary Computation (ICEC 1997), pp. 355–360. IEEE Press, Los Alamitos (1997)

227. Vose, M.D.: Generalizing the notion of schema in genetic algorithms. Artificial Intelligence 50(3), 385–396 (1991)
228. Wang, Y., Wineberg, M.: Estimation of evolvability genetic algorithm and dynamic environments. Genetic Programming and Evolvable Machines 7(4), 355–382 (2006)
229. Wagner, N., Michalewicz, Z., Khouja, M., McGregor, R.R.: Time Series Forecasting for Dynamic Environments: The DyFor Genetic Program Model. IEEE Transactions on Evolutionary Computation 11(4), 433–452 (2007)
230. Whigham, P.A.: Grammatically based genetic programming. In: Proceedings of the workshop on Genetic Programming: from Theory to Real World Applications, Tahoe City, California, USA, July 9, 1995, pp. 33–41 (1995)
231. Wildt, D.E., Bush, M., Goodrowe, K.L., Packer, C., Pusey, A.E., Browne, J.L., Joslin, P., O'Brien, S.J.: Reproductive and genetic consequences of founding isolated lion populations. Nature 329, 328–331 (2007)
232. Wong, M.L., Leung, K.S.: Inductive logic programming using genetic algorithms. In: Advances in Artificial Intelligence – Theory and Applications II, I.I.A.S., Ontario Canada, pp. 119–124 (1994)
233. Wong, M.L., Leung, K.S.: Applying logic grammars to induce sub-functions in genetic programming. In: Proceedings of IEEE Conference on Evolutionary Computation, vol. 2, pp. 737–740. IEEE Computer Society Press, Los Alamitos (1995)
234. Wright, S.: The roles of mutation, inbreeding, crossbreeeding and selection in Evolution. In: Provine, W.B. (ed.) Sewall Wright Evolution: Selected Papers, pp. 162–164. University of Chicago Press (1986)
235. Yang, S.: Non-stationary problem optimization using the primal-dual genetic algorithm. In: Proceedings of the Congress on Evolutionary Computation (CEC 2003), vol. 3, pp. 2246–2253. IEEE Press, Los Alamitos (2003)
236. Yang, S.: Constructing dynamic test environments for genetic algorithms based on problem difficulty. In: Proceedings of the Conference on Evolutionary Computation (CEC 2004), vol. 2, pp. 1262–1269. IEEE Press, Los Alamitos (2004)
237. Yang, S.: Memory-based immigrants for genetic algorithms in dynamic environments. In: Proceedings of the Genetic and Evolutionary Computation Conference (GECCO 2005), pp. 1115–1122. ACM Press, New York (2005)
238. Yang, S.: Memory-enhanced univariate marginal distribution algorithms for dynamic optimization problems. In: Proceedings of the Congress on Evolutionary Computation (CEC 2005), vol. 3, pp. 2560–2567. IEEE Press, Los Alamitos (2005)
239. Yang, S.: Population-based incremental learning with memory scheme for changing environments. In: Proceedings of the Genetic and Evolutionary Computation Conference (GECCO 2005), pp. 711–718. ACM Press, New York (2005)
240. Yang, S.: Associative memory scheme for genetic algorithms in dynamic environments. In: Rothlauf, F., Branke, J., Cagnoni, S., Costa, E., Cotta, C., Drechsler, R., Lutton, E., Machado, P., Moore, J.H., Romero, J., Smith, G.D., Squillero, G., Takagi, H. (eds.) EvoWorkshops 2006. LNCS, vol. 3907, pp. 788–799. Springer, Heidelberg (2006)

241. Yan, W., Clack, C.D.: Behavioural GP diversity for dynamic environments: an application in hedge fund investment. In: Proceedings of GECCO 2006 Genetic and Evolutionary Computation Conference, Seattle Washington, pp. 1817–1824. ACM Press, New York (2006)
242. Yang, S., Ong, Y.-S., Jin, Y.: Special Issue on Evolutionary Computation in Dynamic and Uncertain Environments. Genetic Programming and Evolvable Machines 7(4) (2006)
243. Kim, Y.-i., Kim, L., Cho, L.-K.: Winner take all strategy for a Diploid Genetic Algorithm. In: Kim, J.-H., Furuhashi, T., Yao, X. (eds.) SEAL 1996. LNCS, vol. 1285. Springer, Heidelberg (1997)

Index

Kolibri 360 GmbH
Ferdinand-Jühlke-Straße 7
99095 Erfurt, Deutschland
produktsicherheit@kolibri360.de